Shattered Nerves

Nerves

How Science
Is Solving
Modern Medicine's
Most Perplexing Problem

VICTOR D. CHASE

THE JOHNS HOPKINS UNIVERSITY PRESS
Baltimore

The Johns Hopkins University Press
2715 North Charles Street
Baltimore, Maryland 21218-4363
www.press.jhu.edu

Library of Congress Cataloging-in-Publication Data
Chase, Victor D., 1942–
    Shattered nerves : how science is solving modern medicine's most perplexing problem /
Victor D. Chase.
        p. cm.
    Includes bibliographical references and index.
    ISBN 0-8018-8514-0 (hardcover : alk. paper)
    1. Neural stimulation. 2. Myoelectric prostheses. 3. Implants, Artificial. I. Title.
RC350.N48C43 2007
616.8—dc22   2006009626

A catalog record for this book is available from the British Library.

To my wife Sharon

Then the eyes of the blind shall be opened,

and the ears of the deaf shall be unstopped.

Then shall the lame man leap as an hart,

and the tongue of the dumb sing.

ISAIAH, 35:5–6

Contents

Acknowledgments   xi

Introduction   1

1 / Learning to Listen All Over Again   8

2 / The Body Electric   28

3 / Of Frogs' Legs and Transistors   34

4 / The Grandfather of Neural Prostheses   55

5 / Accidental Pioneers   70

6 / Giving a Hand   86

7 / Looking Back at an Empty Wheelchair   98

8 / The Dirty Little Secret   117

9 / Sound in the Brain   128

10 / In the Eye of the Beholder   152

11 / Nerves of Platinum and Iridium   183

12 / Pins and Needles in the Brain   210

13 / From the Inside Out   226

14 / Reaching the Depth of Depression   236

15 / A Hole in the Center of the Brain   244

16 / Ethics   251

17 / Biomimetic and Superhuman   273

Selected Bibliography   279

Index   281

Acknowledgments

I owe a deep debt of gratitude to numerous patients and researchers who gave unstintingly of their time to share their research, personal experiences, excitement, successes, and disappointments relative to their involvement in the highly promising field of neural prosthetics. The patients, on whom many of the experimental systems discussed in this book have been and are being tested, provided me with immeasurable inspiration. They have quite literally given themselves up to the research and to a person have demonstrated true nerve, altruism, optimism, and even cheerfulness in circumstances that would bring many a lesser person to despair. They spoke of their circumstances, their emotions, and the research they have volunteered to participate in with complete candor and honesty, something I greatly respect and appreciate. Those volunteers I would like to thank are James W. Jatich, Jennifer S. French, Harold Churchey, Connie Schoeman, Marilyn Davidson, Molly Brown, Holly S. Koester, Ronnie Rainge, Maria Zaccaro, Scott Hamel, and Ryan McLeod. And a special thanks to the late Michael Pierschalla, who provided a wealth of information and inspiration and who died far before his time.

Similarly, the many physicians, engineers, and scientists I have had the privilege of speaking and meeting with during my research for this book have impressed me as sincerely committed individuals who care more about their work and the good it can do for the disabled than the greater amount of money they could likely make otherwise, something rare and admirable. I would like to especially thank F. Terry Hambrecht, who, by dint of his own drive and intellect, provided much of the impetus to advance the field of neural prostheses to the point it is today. Terry, who co-founded and headed the National Institutes of Health's Neural Prosthesis Program until his 1999 retirement, gave generously of his time and knowledge during my years of research for this book and was kind enough to review and critique the manuscript. Others deserving of special mention for spending many hours being interviewed by me, either in person or over the telephone, and in some cases both, and for responding promptly and thoughtfully to my follow-up queries, are William Heetderks, who was Terry Hambrecht's deputy and then succeeded him as director of the NIH Neural Prosthesis Program; Robert V. Shannon of the House Ear Institute; P. Hunter Peckham, J. Thomas Mortimer, Graham Creasey, Ronald J. Triolo, Kevin Kilgore, and Dustin J. Tyler, all of the Cleveland Functional Electrical Stimulation Center; Giles Brindley, a British physician and inventor; John P. Donoghue of Brown University and Cyberkinetics; Richard A. Normann of the University of Utah; Donald K. Eddington of the Massachusetts Eye and Ear Infirmary; Douglas B. McCreery of the Huntington Medical Research Institutes; Jonathan R. Wolpaw of the New York State Department of Health's Wadsworth Center; Philip R. Troyk and Michael Davis, of the Illinois Institute of Technology; Mark Humayun, Gerald E. Loeb, and Theodore W. Berger of the University of Southern California and the Biomimetic Microelectronic Systems Center; Robert Greenberg of Second Sight; and Richard A. Andersen of the California Institute of Technology.

I would also like to thank Arthur Caplan of the Center of Bioethics at the University of Pennsylvania; Thomas H. Murray and Mary Ann Baily of the Hastings Center; Joe Schulman of the Alfred E. Mann Foundation; Kensall D. Wise of the University of Michigan; Alan Chow of Optobionics; Mary Buckett of the Cleveland Functional Electrical Stimulation Center; Philip R. Kennedy of Emory University and Neural Signals, Inc.; Andrew Schwartz of the University of Pittsburgh School of Medicine; Sharona Hoffman of the Case Western Reserve University School of Law; Miguel Nicolelis of the Center for Neuroengineering, Duke University Medical Center; and Timothy R. Surgenor of Cyberkinetics.

A special thanks goes to the editorial staff at Johns Hopkins University Press, who, to a person, were a delight to work with. I would like to especially recognize the efforts of my editor Vincent J. Burke, who stood four-square with me on the road to publication of this volume. Hugh Cahill did an excellent job correcting technical errors, and Michael Baker tightened my prose without altering my voice. Both Cahill and Baker used a light touch and saved me from some embarrassing errors.

My heartfelt gratitude goes to Brandon and Sarah for the books, to Abbey for the conversation and artistic input, to Scott for the assistance, and to Alexa, Gabriel, and Sydney for never failing to make me smile. I am also greatly indebted to Cyril R. McGinnis.

I would also like to apologize to those pioneers, researchers, and patient volunteers in the field of neural prostheses, past and present, who have made and are making significant contributions but whom I have not mentioned in this book. By its very nature, this book is intended to provide an overview of a vast and growing field of endeavor and, therefore, cannot be inclusive of each and every individual's contributions, significant though they may be. To do so would require an encyclopedia, rather than a single volume. My goal is to provide you, the reader, with a solid understanding and appreciation of the entire field of neural prosthetics, its people and technology in an interesting and informative format. Once inspired, the reader may wish to refer to the selected bibliography for further reading.

Shattered
Nerves

Introduction

The marvel of the human machine unavoidably inspires awe. The coordination within the massive complex of organs that make up our bodies is nothing short of miraculous. While each organ performs its individual function, it also operates in finely tuned concert with the other instruments of the body to create the music of life. The nervous system alone, consisting of billions of neurons, or nerve cells, that allow us to perceive and interact with the world around us, makes the finest of humankind's technological developments pale into insignificance.

Even scientists who devote their entire lives to understanding the workings of the sensory systems eventually arrive at a gap they've been unable to bridge short of taking a leap of faith. Modern technology allows them to watch an individual's brain waves fluctuate in response to a stimulus such as sound, light, or a pinprick. But they still can't look at a spike in waves on an oscilloscope or changes in images on a brain scan and really understand how that stimulus translates into perception. They don't understand how electrical activity in the brain corresponds to perception, pain, pleasure, or conscious awareness.

There's no doubt, however, that electricity is at the root of it all. Electricity, or the movement of electrons and ions, is such a fundamental aspect of nature that it was woven into the fabric of life. A long time before humankind ever walked the face of the earth, let alone thought about electronics, Mother Nature found that electrical signals provide the most efficient method of transmitting information within the body. No living creature could survive without electricity, because the body is, in essence, an electrical machine. Without electricity, neurons could not communicate the signals that allow us to see, hear, touch, smell, taste, and move about, and even think. We need electricity to interact with the world around us as much as an electric motor requires electric power to function. Without it the motor is dead. The same holds true for human beings. Without electricity there is no life.

Complete comprehension of how small spikes of electricity lead to perception and thought still lurks somewhere in the future. But scientists are making exponential leaps in understanding the mass of neurons that make up the brain and the rest of the nervous system that extends from it, though their task is akin to counting, categorizing, and understanding the activity of each star in the universe, as well as its relationship to the whole. Given this level of complexity, resulting from the vast number of elements that must operate perfectly to provide perception, movement, and thought, it is amazing that it is not the norm for things to go awry. Yet in the vast majority of people, the staggering number of components that make up the bodily systems that allow us to function in our environment work perfectly, or close to it.

Unfortunately, in some people, the circuitry that generates and conducts electrical signals goes bad, rendering them unable to fully partake of the miracle of the senses, as in the case of the blind, when the rod and cone photoreceptors inside the eye can no longer translate light into the electrical signals that send information to the brain. Or when the hair cells inside the cochlea of the inner ear, which process sound waves, die off, and a person loses the ability to hear. Failure of the body's electrical circuitry is also responsible for paralysis that occurs when spinal cord injuries damage the nerve cells that carry electrical signals from the brain's motor cortex to the muscles and from the skin's tactile receptors to the somatosensory portion of the brain. Until recently, these conditions were deemed irreversible. Now there is hope.

Through the ability to miniaturize integrated electronic circuitry, scientists can take concrete steps toward countering the ravages wrought on those whose internal circuitry has shorted out, without it being a total act of hubris. The same methods used to shrink electronic components down to pocket computer and digital watch size are now being used to create reliable, intricate devices small enough to be implanted inside the eye, the ear, the muscles, and the brain itself. These manmade, implantable marvels of modern technology are known as "neural prostheses," devices that directly interface with some component of the nervous system. They do so either by feeding electrical impulses into nerves or muscles or by recording signals from the nervous system and using those signals to operate some kind of machine, which itself may be implanted in the body.

Neural prostheses have the potential to aid the hundreds of thousands, or perhaps even millions, of individuals with neurological disorders that disrupt their ability to move or to communicate. These people have functioning brains, but because of injury or disease, cannot get the output of their brains to the parts of their bodies that should receive the signals or cannot receive impulses to their brains that would enable them to utilize the sensory-processing portions of their brains. Though the idea of mating neural prostheses to the body has been around for quite some time and a number of early researchers did experiments in the field, it is only relatively recently that scientists have had the knowledge of brain function and the technological arsenal to actually create viable neural prostheses.

———

THE FIRST WIDELY USED NEURAL PROSTHESIS to be added to the physician's arsenal against sensory deprivation was the cochlear implant, which was first embedded in the inner ears of people with profound deafness in the early 1970s. Since then, tens of thousands of people have had some measure of hearing restored through these devices. Typically, the wearer of a modern cochlear implant who was completely deaf prior to being implanted, can now carry on a relatively normal telephone conversation. The success of cochlear implants helped pave the way for work on retinal implants designed to give at least partial sight to people who are blind and beyond the help of purely medical ministrations. Utilizing electrodes placed directly on the delicate retina inside the eye, retinal implants are intended to replace damaged rod and cone light receptors that are no longer doing their jobs because of diseases such as retinitis pigmentosa and macular degeneration. In addition to feeding signals to the blinded eye and the deaf ear, researchers are designing systems that bypass the primary sensory organs and feed electrical impulses directly into the visual and auditory cortices of the brain to stimulate sight and hearing. Such systems can be used in patients whose eyes and ears cannot process any signals, even those fed in by means of cochlear and retinal implants because the nerves leading from the ears or eyes are too damaged to carry those signals to the brain. In such cases, bypassing these primary sensory organs by feeding signals directly to the brain through electrodes placed on or in the brain may be the answer.

Another family of electronic implants is currently returning hand movement to quadriplegics, and the ability to stand and step to paraplegics. In this facet of neural prostheses, called functional electrical stimulation, or FES, scientists are merging humans and machines by implanting electrodes directly into the muscles of people with paralysis. Computer-controlled jolts of electricity stimulate the muscles causing contraction and movement. This can be achieved because even though one is paralyzed, one's muscles are usually intact despite damage to the nerve pathways that feed signals to them. The first U.S. Food and Drug Administration–approved FES device, appropriately named the Freehand, is giving hundreds of quadriplegics the ability to feed and groom themselves, and in some cases implantees can even operate computers using their hands. Though the Freehand was short-lived as a commercial product, for business rather than technical reasons, its developers are still working to improve the technology. And the FDA-approved Vocare bladder control system uses neural prosthetic technology to return bladder and bowel control to people with paralysis for whom these are major problems. In some cases, the same device produces erections in men.

Scientists are also developing technology that may return the sense of touch to users of FES systems by using electrodes to record signals from a patient's own tactile receptors, which along with muscles, remain functional in spite of paralysis. Early efforts are aimed at improving the grasp capabilities of Freehand users, who have only visual feedback, which does not provide the subtlety of grasp available to the able-bodied. A touch-sensitive neural prosthetic system records signals from the tactile receptors in the user's hand and feeds them directly to the prosthesis's computer, which uses the information to adjust the pressure of the grip. Early systems do not enable the patient to feel what he or she is holding, even though the tactile feedback system uses the body's own sensing apparatus to determine the pressure required to grasp a cup, for example. The hope is to eventually return the sense of touch directly to the patient, initially by remote referral. Pressure on the hand would activate a stimulation device to apply pressure to a part of the body above the severed nerves where natural feeling still exists. The ultimate goal is to record signals from healthy tactile receptors and transmit them to microelectrodes implanted directly in the brain's somato-sensory cortex, where skin, muscle, and joint information is processed.

Another twist on the same theme involves sending signals in the opposite direction. Instead of transmitting them to the brain, electrodes implanted in the motor cortex of the brain—where electrical impulses initiating movement, known as action potentials, are created—can capture intentions, which would then be used to activate FES devices. This would be accomplished by transmitting action potentials recorded by the electrodes in the brain via a computer to electrodes implanted in paralyzed muscles, thereby effectively bypassing the damaged nerves in the spinal cord and allowing wearers to operate devices, such as the Freehand, merely by thinking about it, much as able-bodied people move their limbs. This differs significantly from the current configuration, in which the Freehand is operated by a joystick mechanism mounted on a part of the body unaffected by paralysis, such as the shoulder. The same brain implants that send signals to the Freehand via thought could also be used to operate a robotic arm that would respond to the wishes of the patient, or to operate a wheelchair. This technology can also give "locked-in" patients, who can neither move nor speak because of stroke or disease, such as advanced amyotrophic lateral sclerosis (ALS, also known as Lou Gehrig's disease), the ability to communicate. Though many such patients remain intellectually astute, they find themselves in one of the most fearful dilemmas a human can confront—being totally sentient yet unable to communicate with anyone. With the neural prosthetic technology that records their thoughts and allows them to essentially think a computer into operation, bypassing the need to move a mouse or type on a keyboard, locked-in patients can again interact with their fellow human beings.

The same technology that can enhance the lives of people with severe disabilities also holds the potential to expand the capabilities of the able-bodied with as yet undreamed of consequences. The visible wavelength may be increased, or the ability to hear sounds that only animals with more sensitive ears can now perceive may fall within human capability. And learning capacity and memory may be increased. The U.S. Air Force has looked at the technology as a possible way of augmenting the ability of fighter pilots to operate the highly complex systems in their aircraft. And in what sounds like science fiction, but has realistic potential, a leading physician in the study of how the brain represents tactile information says he believes the brain is capable of incorporating a machine into its representation of the body. In other words, from a sensory standpoint, an autonomous machine could be made part of a person. The individual would experience the same sensations as the machine. This would, for example, enable an earthbound scientist to explore another planet by seeing and feeling what a robot actually located on that planet perceives. By the same token, a safely ensconced individual could have a machine do all sorts of nefarious deeds on his or her behalf, essentially without detection.

Though some of the hopes and goals of the scientists involved in developing neural prosthetic implants may seem far-fetched and perhaps impossible, experience has shown that if it can be conceived it can be done, given time, money, and the tools made available by modern technology. Consider, for example, the 300-year-old drawing by Isaac Newton of a man on a mountaintop throwing a ball into a parabolic arc around the earth. During Newton's time, the idea of putting a manmade satellite into orbit around our planet would have undoubtedly been considered the musings of a madman, yet, though it took hundreds of years, Newton's dream is today a reality.

It is, therefore, an extremely exciting time for those working in the field of neural prostheses as well as for those who may benefit from the fruits of their labors. But a word of caution is prudent. As is the case with any emerging field that holds great promise, overzealousness on the part of some of those involved can result. Thus, people with disabilities who may, in fact, someday be aided by developments in this new area of technology should not allow their hopes to get unrealistically high. Blind individuals, for example, should not expect full sight restoration, but instead perhaps the ability to see only points of light or shadows that may enhance mobility. And people with paralysis cannot expect to stand and walk with a normal gait and without the assistance of a walker anytime soon. Most researchers in the field themselves expect only relatively modest gains in the short term. Neural prosthetic technology does indeed hold the promise of returning almost normal functioning to those whose nervous systems are impaired, but that remains a hope for future generations. In the meantime, the step-by-step gains will likely be more modest. Yet as virtually every person who has volunteered as a test subject for the research and development currently being conducted has said, "Something is better than nothing."

But even if some of the loftiest goals of this work are never achieved, the act of striving toward them will not be for naught. For it is certain that neural prosthetic research—especially the facet of it pertaining to brain implants—will go a long way toward solving the mysteries of some of biological science's last frontiers, such as how the brain and sensory systems function. Through the electrodes implanted in the brains of human patients, scientists for the first time have an unobstructed view into the workings of the brain. "You can tell someone to imagine something, but you can't tell a monkey to do that," said Andrew Schwartz, a leading researcher in the field, at the University of Pittsburgh. "Through the use of language and comprehension you can do all sorts of experiments that you could never dream of before, and the data we could get would be very rich." Through such work, scientists might finally be able to understand how perception gets transformed into consciousness and how a pinprick actually does make you say ouch. In the meantime, the quest for better neural prostheses goes on, and though no one is claiming to be anywhere near bionic nirvana, the pursuit is indeed electrifying.

1

Learning to Listen
All Over Again

Michael Pierschalla, an extremely smart, sensitive individual, grew up in the small central Wisconsin city of Wausau. In the autumn of 1974, at the age of 19, he moved 140 miles south to attend the University of Wisconsin at Madison, not knowing exactly what he wanted to do with the rest of his life. Like a lot of disaffected young people during the Vietnam War era, Pierschalla had an unfocused thirst for knowledge, which led him to study philosophy and spend time with his friends drinking coffee, smoking cigarettes, and playing his guitar while trying to figure it all out.

About six months into his freshman year, he decided he wasn't ready for college. He felt he was wasting both his time and his parents' money, so he packed his bags and returned to Wausau, where he moved into the basement of his parents' home and took on odd jobs. The focus of his life was his evenings spent at the BonTon Café with those friends who had not yet left town, attempting to solve the world's problems. "There were a bunch of us just sort of waiting for our calling. This was pretty much how the days passed, one after the other," said Pierschalla. He spent the rest of his time involved with his first love, music. Headset on, and guitar in hand, he taught himself to read music and dreamed of becoming a world-famous guitarist. "My very earliest memories are essentially sound-oriented ones. Music was really something that moved me very deeply," he said.

Then, in early August—on a Thursday, to be precise—Pierschalla's life started to unravel. "I remember staying out very late the night before. I then walked a couple of miles back to my folks' house, and went to sleep. I have a recollection of waking up in the middle of the night and turning over in bed and feeling a little bit of an odd sensation, but I'm not sure just what it was. When I did wake up in the morning a little bit on the late side, I got up out of bed and stumbled and fell over." He also heard a ringing in his right ear. No one else was home at the time, so he looked up the address of the closest ear, nose, and throat (ENT) specialist and unsteadily walked several miles to his office. "They gave me an ear exam with an otoscope and said they didn't see anything unusual. That I shouldn't worry about it too much and that the regular doctor was out of town for the weekend and would be back on Monday and I should come back then." Pierschalla was not appeased. "I had an intuition that something was seriously wrong."

As time progressed, his condition continued to deteriorate. The ringing got louder in his right ear, and he started to hear buzzing in his left ear. Vertigo and nausea set in. On Friday, he paid a visit to the family doctor, who gave him an antivertigo medication and a tranquilizer to keep him calm until the ENT physician returned on Monday. In the meantime, Pierschalla's hearing deteriorated to the point he was having a hard time understanding people. Trying to keep his equilibrium, he went to the house of a friend who was a jazz aficionado and who put on a recording of a Norwegian trio. "I wasn't hearing it very well, but I seem to recall that was probably the last album that I heard with much fidelity to it," said Pierschalla.

By Saturday, he was having difficulty walking and experienced loud ringing in both ears. Striving to maintain a semblance of normalcy, he tried walking to the café where he and his friends regularly gathered. "I was stumbling down the street as if I was very drunk . . . At one point I remember stumbling and falling into bushes on the side of the road, but I was determined to make it down there." By Sunday, Pierschalla and his parents decided to wait no longer and went to the local hospital. Except for his symptoms, Michael was in good health. He had no fever, no obvious infection, no history of hearing problems, and he had no illness immediately before the onset of his bizarre symptoms. Every test that was run came back negative, leaving his physicians completely baffled. After three days of testing with no conclusive results, he was discharged. Though distraught, Pierschalla maintained hope and faith in medicine. "For heaven's sake," he recalled thinking, "they can open up somebody's skull and dig out a tumor, they can replace hearts, somebody is going to know what this is. It's just that I'm living in a small town. Somebody's going to find out what this is and they're going to fix it."

In search of that fix, Pierschalla went to another clinic—about a month after the onset of his symptoms, which remained with him—where he was put through another battery of tests. He recalled a great sense of frustration when he returned with his parents and the three of them sat down to talk to a doctor who seemed to ignore him and directed his conversation to his parents. The physician then met privately with Michael's parents, and when Pierschalla was called into the room his mother was crying. "The doctor told me, 'You've lost your hearing. It's not the kind of thing that comes back. You are going to have to learn to live without your hearing. Many people do.' "

His hope shattered, he went to pieces. "My despair was about as great as it could be. At that point I had sort of a major breakdown. They had to tranquilize me for quite awhile. I had no idea what to do next. More than anything I guess I was filled with a lot of fear," he said. Still not knowing what was wrong with him, doctors gave Pierschalla a course of steroid therapy, hoping to alleviate a possible inflammatory reaction in an effort to preserve any chance that some hearing might return. Because he was unable to hear anything, family and friends wrote notes to Michael, and he would reply orally. One of his biggest sources of despair was that he could no longer enjoy his beloved music, "despite the fact it was always going on inside my head. There was always a radio playing inside my mind," he said.

By Christmas of that year, Michael began to come out of his depression and decided to get his life on track. "I realized that I was facing a long future without my hearing, that the best I could hope for was to hear again in my dreams, and I made a vow never to forget the music, the sound of rain falling on the sidewalk, and my parents' voices. And then I went on and faced life in a different way, looking for another identity."

He moved out of his parents' home and took an apartment of his own.

He also enrolled at a local community college, where a note taker was assigned to him. A friend who was an artist urged Pierschalla to try some art classes as an outlet for his pent-up creative energy. He tried drawing and ceramics, but neither of them clicked. What did soothe his suffering was working in his father's basement woodworking shop making furniture for his apartment. "I found it was the kind of thing that kept me somewhat calm and kept me involved a full twelve hours or so a day," said Pierschalla, who also found intellectual fulfillment in taking a fine-art approach to woodworking. As was his nature, he delved deeply into the field. In the course of his studies, he learned of the School of American Crafts at the Rochester Institute of Technology (RIT), where one could study woodworking. For the first time since the onset of his deafness, he found something that truly excited him. "It opened my eyes. There was a place you can actually go to college and earn a degree in working with your hands. So I became very interested in that," he said. Ironically, Pierschalla was unaware at the time of RIT's National Institute for the Deaf, a leading college for students who are deaf or hard of hearing.

While he continued to work in his father's basement shop and contemplate attending RIT, Pierschalla experienced another health-related episode that put him right back on an emotional roller coaster. Nine months after the onset of his original symptoms, his eyes suddenly became inflamed to the point he could hardly open his eyelids. "I freaked out totally because there I was deaf—I couldn't hear anyone—and then suddenly I could hardly open my eyes to see. I was scared shitless to tell you the truth," he said. "I was going deaf and blind at the same time."

Fortunately, a regimen of eye drops cleared up the problem in about a week. When the disease Pierschalla had contracted was correctly diagnosed several years later, it was recognized that eye inflammation is part of the syndrome's pattern. Specifically, he was suffering from Cogan's syndrome—technically, nonsyphilitic interstitial keratitis with sudden onset deafness—which was first identified by David Cogan, an ophthalmologist. Although Cogan first described the symptoms of this extremely rare disease in the mid-1940s, it was not until the first major compilation of case studies was completed during the early 1980s at the Mayo Clinic in Rochester, Minnesota, that Cogan's was defined as a syndrome. Still somewhat of a mystery, the general consensus is that it is an autoimmune disease.

One thing certain about the syndrome Pierschalla experienced was that it significantly damaged the cochleae in both of his ears. The cochlea resides deep inside the head, is less than a half-inch in diameter, and resembles a miniature snail (the name is derived from *kokhlos*, the Greek word for snail). This highly complex organ is the final mechanical processing portion of the auditory system, where vibrations are converted into electrical signals for transmission to the brain. It is one of three components of the inner ear, the other two being the semicircular canals and the vestibule, both of which contribute to equilibrium.

The hearing process begins at the outer ear, which focuses sound waves on the eardrum, also known as the tympanic membrane. It, in turn, vibrates the ossicles, three tiny bones in the middle ear that in evolutionary terms evolved from the jaw: the malleus (hammer), incus (anvil), and stapes (stirrup). These bones mechanically amplify the vibrations approximately twenty times. The stirrup is like a plunger connected to a membrane known as the oval window—the entryway to the fluid-filled cochlea. As the oval window moves, it alters the pressure in the cochlear fluid. This flexes another membrane, called the basilar membrane, which in turn actuates the organ of Corti, a structure containing approximately 12,000 hair cells. These hair cells rub against the tectorial membrane. The resulting deflection of the hair cells causes them to release neurotransmitters to some 30,000 nerve fibers that make up the auditory nerve. The nerve fibers then produce electrical signals that are transmitted to the brain through the auditory nerve.

Virtually all sounds encountered in nature are complex, in that they contain energy in many frequencies. To handle this vast array of frequencies, the cochlea functions like a mechanical spectrum analyzer separating the frequencies and sending them to the appropriate nerve fibers, or frequency channels. Each frequency hops a ride to the brain aboard its own private group of auditory nerve fibers. When it reaches the auditory brainstem, the signal is split into several pathways, where particular aspects of the acoustical signal are analyzed. There are, for example, specific structures in the brain that compare the sounds coming in from both ears to identify the location of that sound in the horizontal dimension. Another area determines the vertical position of the sound. Yet other pathways extract information about the spectral shape of sounds, such as the cutoffs between one frequency region and another.

In about 99.9 percent of the people who lose their hearing, it is the death of the hair cells that causes deafness. Diseases that damage the hair cells in the cochlea frequently damage the vestibular portion of the cochlea as well, leading to a loss of the sense of balance along with a hearing loss, which is what happened to Michael Pierschalla. One of the many manifestations of a loss of balance is an inability to keep things in focus while one is moving around, when riding in a car, for instance. Pierschalla described the sensation as being similar to watching an earthquake as filmed by a shaking camera. With his vestibular system inoperative, Pierschalla was also not able to feel himself moving without visual cues. Typically, if you close your eyes while swinging in a hammock, you can feel yourself move. An individual whose vestibular system does not function properly does not get that same sense of motion. Similarly, on amusement park rides, "on a Tilt-a-Whirl, for instance, if I close my eyes I don't have a sensation of going around in circles," said Pierschalla. As a result, he didn't get motion sickness, which would be advantageous for an astronaut but not for someone trying to navigate his way around terra firma.

Pierschalla was able to compensate for the vestibular loss by using other senses, including proprioception—the sense that relies on sensors in the muscles to keep one apprised of where he is in space—combined with his visual perspective, skin sensors (especially those on the soles of his feet), and joint angle sensors. To improve his ability to keep his balance, Pierschalla taught himself Tai Chi, a Chinese martial art form that combines yoga and meditation, which he said, "is all about centering yourself in the earth, and gaining a sense of balance that's not just oriented in your vestibular system, but in your whole body." Pierschalla's perseverance, discipline, and sheer willpower were such that even without a sense of balance, he was able to retrain himself to ride a bicycle. The first time he tried, "I went about two pedal strokes and fell right over," he said. But he stuck with it and learned. It was in the dark, when there was no visual horizon to lock onto, that balance problems became insurmountable. He told of "one instance when I had gone with some friends for a cookout. We stayed late and sat around a fire until it got dark. For me trying to walk back down the path in the dark was virtually impossible. I ended up having to have somebody lead me front and back so I didn't continually fall over."

As Pierschalla was rebuilding his life after years of hardship and depression, something positive finally came his way. Occasionally, he noticed tiny sounds. "I would hear a little click when I slammed the car door," he said. And while chopping and stacking firewood, he heard little snapping sounds as logs dropped. "I began wondering what was going on, so I went into the kitchen and grabbed two of the biggest pots I could find and slammed them together right next to my ear, and I could actually hear a little something." A visit to the audiologist confirmed that about 5 percent of his hearing had returned in his right ear. To make the most of it, he was fitted with the most powerful hearing aid available. The electronics fit in his shirt pocket, and it had a large ear mold that made Pierschalla "profoundly embarrassed. The hearing aid was a symbol of something I couldn't control and couldn't regain and couldn't do anything about. It was a badge of what I had become and it made me sad," he said. Yet this big piece of hardware helped a lot with lipreading, which Pierschalla had learned intuitively. So despite his embarrassment, he used it.

He then decided to apply to RIT's School of American Crafts. But when he did, he was informed that he would have to start out at RIT's National Institute for the Deaf, which was not affiliated with the crafts program. He wasn't happy about the change but decided to attend RIT nonetheless. During a summer orientation program for incoming students, Pierschalla felt "like a fish out of water." Most of the deaf students used sign language, which he was not familiar with, he was older and more experienced than the other freshmen, and he had already selected his vocation. "What I wanted more than anything in the world was to get a spot in the wood shop on the other side of the campus, which had a long waiting list of its own," said Pierschalla. Instead, he was obliged to take sign language training and basic art courses. Through persistence, he was eventually allowed to take some night courses in the woodworking program, and once his talents became apparent, he was given a full-time slot.

During his sophomore year, a friend asked him to join her at the Appalachian Center for Crafts, a division of Tennessee Technological University in Smithville, where the focus was on training craftspeople of all sorts. Excited about the idea, he took a leave of absence from RIT—to which he never returned—and headed for the Smoky Mountains of Tennessee to help establish a woodworking program. In the classes he taught at the Appalachian Center, he dealt with his deafness by telling his students, "You can understand me perfectly, and I can understand you if you're willing to write. Use some creative strategies and there's always the chalkboard. I've been through this before, it's not that hard to communicate, and we'll solve the problems as they come up. I would tell them it's not their fault and it's not my fault so let's just work with things the way they are, and we did just fine."

Pierschalla spent two happy years in Tennessee. Then fate dealt him another blow. When he awoke one morning and put on his hearing aid, he heard nothing. Thinking it was a malfunction, he tried his backup unit and still heard no sound. He also had some strange sensation in his right ear but gave it a day to see if it would stabilize. It didn't. Whereupon he contacted his doctor in Wisconsin, who told him to immediately see a colleague of his in Nashville, some 60 miles from Smithville. When Pierschalla got there, he was hospitalized and given several days of histamine therapy to stabilize his condition. It didn't work, so exploratory surgery was done to find the culprit. None was found, but the prognosis was that Pierschalla's residual hearing had left him for good. He was now totally deaf.

His reaction this time was more resignation than depression. "The first time I lost my hearing I was depressed for two full years. This time I said I don't have that much time to lose again. There's nothing I can do about this. Of course I was sad, but I decided I had to take it one day at a time and keep on with my life. My work was keeping me pretty busy. I had learned to function pretty much as a hearing-impaired person. Now I was going to have to become a deaf person. It didn't substantially change who I was inside, and I kind of went on with my life."

His philosophical bent coupled with his prodigious strength of character and maturity helped him adjust. "When we talk about the loss of one of our senses or abilities, we talk almost exclusively in the language of accommodations, adaptation, and acceptance," he said. "Faced with this sort of challenge, our concept of healing and recovery often takes on a new meaning, one that is focused on a spiritual and psychological, rather than a physical recovery. If we can no longer expect to join nerves back together, we try instead to reconfigure the soul and the self."

Part of his reconfiguring involved seizing a new opportunity. He had been invited to join a woodworking cooperative in Cambridge, Massachusetts, where a group of about fifteen artisans shared equipment and expenses. In 1982, he decided to accept the offer. There Pierschalla built fine furniture on speculation for gallery shows, as well as on commission from people who had seen his work at the galleries and wanted a particular piece. Superb as his work was, he never touted the fact that some of his pieces were later displayed at the Museum of Fine Arts in Boston, the American Craft Museum in New York, and the Smithsonian Institution in Washington, D.C.

Pierschalla had also become an inveterate journal writer. His journal was a friend he could talk to comfortably, whenever and about whatever he liked. It also served as his therapist, providing him a way of thinking through problems. In those days, it was not uncommon to see him sitting in Harvard Square writing journal entries.

Within two years he was as well settled into his new life as possible. "I was living in Cambridge and just getting by, existing as a deaf person," he said, describing himself as, "a hearing person who couldn't hear any more. My voice never changed. I was the same individual, but without my hearing."

Then he came upon an article about a new kind of device being developed called a cochlear implant, which held the promise of returning hearing to the deaf. His reaction was, "Yeah, yeah, yeah, but that's not for me. I'm sensory neural deafened and that can't be changed." But he had friends who thought otherwise and encouraged him to seek out information about the cochlear implant. And he was, after all, a fighter. Although on some days the candle of hope flickered dim, it had not been extinguished.

Cochlear implants are designed to replace the function of damaged hair cells within the cochlea with electrodes that are inserted into the winding cochlea adjacent to the nerve fibers. Electrical impulses sent to the electrodes provide the stimulation of the auditory nerve fibers that would otherwise be activated by healthy hair cells. The research that Pierschalla had read about was being conducted by Donald K. Eddington, who had moved from the University of Utah to the Massachusetts Eye and Ear Infirmary to direct its Cochlear Implant Research Laboratory in 1983.

Eddington was one of several researchers at the time who were striving to bring the cochlear implant from a rudimentary device to a system that could enable deaf individuals to hear speech. Another cochlear implant developer was William House, a surgeon whose older brother Howard House founded the Otologic Medical Group in Los Angeles, which in 1981 became the House Ear Institute. William House fashioned a single-channel device that in the 1970s became the first cochlear implant to be commercialized. Manufactured by 3M and known as the House/3M cochlear implant, it scored several firsts, including being the first neural prosthesis to be implanted in large numbers of adults and the first such device to be implanted in children. Though the single-channel implant did not enable most users to recognize speech, it did help them with lipreading.

Despite the success of the House implant, whether human implantation should continue without further animal testing remained a contentious issue in the field. To put the controversy to rest, the National Institutes of Health Neural Prosthesis Program funded a study of single-channel cochlear implant recipients led by Robert Bilger, of the University of Illinois at Urbana-Champaign. The result was a groundbreaking work, still referred to in the field as the "Bilger Report." It covered all aspects of cochlear implantation, and according to a paper written by F. Terry Hambrecht, a founder of the Neural Prosthesis Program, the report concluded "that for profoundly hearing impaired individuals, the single channel cochlear implant could be a more useful and acceptable cuing device than other traditional devices, such as hearing aids. It substantiated reports that, for some individuals, cochlear implants improved lipreading, provided awareness of environmental sounds, and improved control of voice modulation. This study is considered the turning point in the acceptance of cochlear implants by basic scientists, audiologists, otologists, and the community of profoundly deaf individuals. The report also clearly indicated the need for improvements in cochlear implants which helped propel the development of multichannel cochlear implants."

Spurred on by the Bilger Report, researchers at numerous institutions pursued multichannel cochlear implant developmental work. Eddington was one of those researchers. Working in concert with Derald Brackmann, a physician at the House Ear Institute, Eddington developed one of the first multichannel cochlear implants while he was still a graduate student at the University of Utah, in Salt Lake City. He wrote his dissertation on the cochlear implant and received his doctorate in medical computing and biophysics in 1977.

Eddington's new six-channel device was implanted in four patients with profound deafness. At the time, no portable processor system existed, so the patients had to go to Eddington's laboratory at the university to be hooked up to an experimental processor that fed signals to their implants. So that their hopes were not falsely raised, the test subjects were informed prior to agreeing to be implanted that their implants were purely experimental and that the goal of the tests was to answer numerous questions, such as how many electrodes were needed to create functional sound? How much current had to be used to elicit such sounds? How should the input be processed to realize optimum use of the electrodes? And would electrodes placed at various locations along the cochlea elicit different pitches? The volunteers were also informed that their implants would likely be of no practical use to them and that they would probably never receive portable processors. When first hooked up to sound-processing equipment in the laboratory, the volunteers were able to discern only beeps and buzzes. Yet even these noises had a strong emotional impact on the patients. One individual told Eddington that each time he had to leave the laboratory, he felt as though he was losing his hearing all over again.

As Eddington and his colleagues tested and learned which frequencies and electrode combinations worked best, they were able to provide more sophisticated input to the electrodes, which enabled the test subjects to discern an ever increasing range of sounds with their implants. Eventually, as the researchers came to better understand how to present electronic information to the implantees, all but one, who had been deaf since birth, were able to recognize some speech without lipreading. Eddington's group was able to improve the sound the implantees heard by taking advantage of the fact that the cochlea is tonotopically arranged much like a piano keyboard, with the low-pitched notes at one end and the high-pitched notes at the other. High frequencies (in the 20,000 hertz range) are processed by hair cells at the entry to the snail-shaped spiral of the cochlea, with the frequencies becoming progressively lower the deeper the cells are located in the winding structure, until they reach the lowest audible frequency of some 20 hertz. This neat tonotopic structure enables the programming of the electrodes positioned throughout the cochlea to provide the auditory nerve with stimulation at the points at which it would normally receive input from various frequencies.

Significant as these achievements were, Eddington decided that to progress further with his work he needed a stronger grounding in auditory physiology, so he moved to the Massachusetts Institute of Technology and the Massachusetts Eye and Ear Infirmary to pursue postdoctoral work in the field. He spent two years there and would have stayed permanently, but despite having advised his Utah test subjects they would not have portable sound processors, he felt an obligation to "get those individuals something they could wear." So he returned to Utah for three more years, during which he designed and built a portable processor for the multichannel implant.

The University of Utah subsequently licensed Eddington's cochlear implant design to Symbion, a Salt Lake City company that manufactured and sold the system under the Ineraid name. One drawback to the Ineraid system was that it used a connector that protruded through the skull of the user to connect to the external sound processor. The advent of newer systems that used transcutaneous stimulation, in which signals are transmitted through the skin, precluding the need for a connection protruding through the scalp, led to the eventual demise of the Ineraid system. Yet before it went out of production, the Ineraid returned some measure of hearing to many deaf people and helped substantially further the science of cochlear implants.

Having fulfilled the obligation he felt to his Utah patients, Eddington returned to Cambridge in 1983 to become director of the Cochlear Implant Research Laboratory at the Massachusetts Eye and Ear Infirmary as well as a member of the faculty at Harvard Medical School. It was while seeking additional patients to work with to improve the cochlear implant that Eddington met Michael Pierschalla.

When a friend of Pierschalla's got in touch with Eddington, he was asked to have Michael send in his medical records. Figuring he had nothing to lose, Pierschalla sent them to Eddington and asked whether the work being done was applicable to his problem. Pierschalla was fully expecting to be told no. Instead, he was informed that he would be a very good candidate. Following several meetings with Eddington, Pierschalla was selected to be one of five subjects for the experimental cochlear implant work.

While Eddington now had the capability of offering his test subjects portable devices to use in conjunction with their implants, he still made a point of not building up their hopes. "We were told at the time it would never be like normal sound, it's not like natural hearing," said Pierschalla. He decided to proceed anyway. "I knew virtually nothing about implants, but it was obvious from some of the papers Don wrote that it was doing something, because they were aiding people in lipreading. I thought, 'It's an experiment, but anything is better than nothing.'" He also had an inkling that "something was going to change, and it was going to be dramatic for me, regardless whether it worked out or whether it didn't."

The Ineraid system Pierschalla was to receive included six electrodes, although at the outset only four would be turned on. One of Eddington's goals in working with Pierschalla and the other volunteers was to develop an electronic processing system that would optimize the capability of the multi-channel device so that wearers would eventually be able to more clearly discern speech. Toward that end, Pierschalla received his implant in June 1985. Though he spent only one night in the hospital, he had to wait three months to allow for healing and stabilization before being hooked up to processing equipment in Eddington's laboratory. In an indication of how far the field has come since then, current cochlear implant recipients receive their processors a few weeks after surgery.

When the electrodes in Pierschalla's cochlea were finally turned on and fed computer-generated signals, the first sounds he heard—after years of silence—were beeps, clicks, and buzzes. Despite the fact that the sounds were not coherent, that he was able to hear anything was a highly emotional experience for him. After three months of laboratory testing and six months after he was implanted, Pierschalla was fitted with a portable processor designed to translate sound from the real world into electrical impulses to activate his electrodes. When it was first turned on in Eddington's laboratory, Michael Pierschalla was rendered speechless. "It was hard having the sensation of some sound going on in my head in response to my voice," he said. And as he went forth into the world with his new implant, he was ecstatic. He strutted the streets a proud man. While the hearing aid he had worn when he first went deaf was a symbol of something he couldn't control, the implant was a badge of honor. "It became a symbol of all the new things I could do, even though it was physically much bigger than the hearing aid," he said. He even found himself disappointed when people didn't stop him on the street to ask, "What's with the industrial-sized hearing box?"

"I was sort of floating on air for about a week. It took a while for my feet to touch the ground," said Pierschalla. After years of deafness and depression, he was now like a kid in a candy store. In this case, the ear candy was sound. He went around tapping on walls, shuffling papers, and doing whatever else he could to create sound and to listen and learn.

Although the sounds Pierschalla heard with his cochlear implant were not natural, "it was obvious within the first few days that there was a lot more information there than anyone would have predicted," he said. Describing the sound he heard as "thick and muddy" he could, nonetheless, tell the difference between a male and female voice, and he was able to "hear a lot more in terms of pitch dynamics than what I expected. We expected it to sound monotone. In my case it didn't sound monotone at all. It just didn't have the crispness, or sharpness or fidelity that you think of with typical audition." He also noted the lack of a sense of location. "I don't know where sound is coming from unless I'm visually connected to its source," he said.

By the third day with his new processor, Pierschalla managed to put a telephone call through to his family in Wisconsin, though he was too excited to recall how much he was able to understand of the conversation. Yet many years later, with the original hardware in his head, and greatly improved software, Pierschalla spoke of the early days with his cochlear implant during a clear and comfortable telephone conversation with this author, whose voice he had never heard before. He was also able to enjoy his beloved music again.

—⁓⁓⁓—

WITHIN THIRTY YEARS OF THEIR INCEPTION, cochlear implants evolved from one- and two-channel systems to devices that include twelve to twenty-two channels, enabling a wider range of frequency input—still a far cry from the 3,000 independent pieces of information that the normal cochlea sends to the 30,000 auditory nerve fibers. When cochlear implants were undergoing development, no one knew how many electrodes would be required to provide enough signals to enable the brain to interpret the impulses it was receiving, or even if any sound would be heard. Some scientists thought thousands of electrodes would be required. That even the input from a single electrode was enough to enable a deaf person to hear some sound, albeit far from normal, is a testament to the adaptability of the human brain, which can take such rudimentary information and make sense out of it.

This point was made clear by a computer program developed by Robert V. Shannon, director of the Department of Auditory Implants and Perception Research at the House Ear Institute. Shannon's program allows people with normal hearing to hear what a cochlear implantee hears. To accomplish this, he broke the sound spectrum discernable to the normal human ear into a few segments representative of the number of electrodes an implant contains. This causes a loss of many of the subtle distinctions normally heard. He then filled the frequency bands with noise, essentially dumbing down the ear. The accuracy of the simulation was verified by testing it on people with normal hearing, who made the same vowel-, consonant-, and sentence-recognition errors as do cochlear implant wearers.

When Shannon played the simulation of himself reading aloud as it would sound through a single-channel implant, one could hear the tempo of the speech, but it was unintelligible. Two channels were almost as bad. But when he played the simulation using four channels, the quality of which was also quite poor, an unimpaired listener was able to understand every word. This was because the brain fills in the substantial blanks. The automatic and subconscious nature of this phenomenon is verified by the fact that people who speak English as a second language, even though they may do so fluently, frequently cannot decipher the four-channel message spoken in English. Yet cochlear implant users with as few as one or two channels find that the prosthesis greatly enhances their ability to lip-read. In fact, these individuals frequently think they are hearing what is being said. It is only when not looking at a speaker that they realize they cannot really understand what they hear without lipreading.

Though a variety of modern, commercial cochlear implants differ somewhat, their basic makeup is the same. The electrodes are buried in what looks like thin nylon fishing line that is coiled at one end so it resembles a tiny cane with a convoluted handle. The self-flexing coil is inserted into the cochlea and connected by a set of wires to a receiver-stimulator, about the diameter of a quarter and twice as thick, that is embedded in the skull just behind the implantee's ear. The receiver package also contains a small magnet. The skin over the receiver is completely closed so there is no direct connection between it and the transmitter coil, which also contains a magnet. The attraction of the magnets causes the transmitter to pop onto the skull directly above the receiver, as a magnet latches onto a refrigerator door. The transmitter is connected to a microphone-processor unit that has been reduced in size so that the entire apparatus is no bigger than a hearing aid that hooks over the ear and is barely noticeable. The microphone picks up sound and relays it to the processor, which translates it into radio frequency signals that are transmitted to the embedded receiver-stimulator, which in turn sends them on to the electrodes in the cochlea as stimulation pulses. Modern cochlear implants are so successful that thousands of people who have received them—many of whom could not have heard a shotgun fired next to them prior to implantation—can now speak on the telephone with relative ease. One researcher recalled a cochlear implant recipient who became a telemarketer, and wondered aloud whether "we've done the world a good thing."

—⊸⊸◦∿∿◦⊶⊶—

THE FACT THAT MICHAEL PIERSCHALLA became one of the most outstanding early cochlear implant success stories with only six electrodes in his cochlea is due in no small measure to his intelligence and perseverance and also to ongoing improvements in signal processing that enabled ever increasing amounts of information to be fed to the same six electrodes. This was all accomplished through the external processor he carried in a makeshift shoulder holster and that he referred to as "the mini-computer that is the heart and soul of the implant." Thanks to these signal-processing improvements, "What I hear now has a lot more clarity, sharpness, fidelity, pitch discrimination, and range," he said several years after being implanted.

In fact, as a highly articulate and frequently sought-after test subject, Pierschalla had a lot to do with the substantial improvements that have been made in processor technology. His innate curiosity led him to read voluminously about the science of sound, which in his own words, made him "adept at describing subtle differences from one change to another as we would try different things in the lab."

"He was the ideal subject because he was very bright and he was a very good user so you could quickly change the characteristics of his implant and he could tell you if it was better or worse. He was an important contributor to the development of the new coding strategies in multichannel implants," said Terry Hambrecht, whose NIH Neural Prosthesis Program has provided much of the funding for cochlear implant development. The coding strategies Hambrecht referred to are the various software-based methods of presenting information to the electrodes embedded in the cochlea. Improving these codes so that the signals sent to implanted electrodes provide as much information as possible is one of the major areas of ongoing developmental work in the field.

The earliest cochlear implants operated on an analog processing system in which the electrical signals sent to the electrodes were analogous to the sounds in the environment, just as an amplifier sends continuous, fluctuating amplitude waveforms to high-fidelity speakers. This caused current to flow simultaneously to multiple electrodes, sometimes resulting in excessively loud or distorted sound. Two types of sound-processing technology were subsequently developed to deal with this spillover problem.

One technique is known simply as SPEAK, for spectral peak signal, while the second is called the continuous interleaved sampling, or CIS. Both are digital systems; they break sound down into frequency bands, much like the bouncing bars on a stereo equalizer as it processes musical sounds. Those bars, or frequency ranges, are processed and presented to the electrodes sequentially, not continuously, so that no two electrodes are actuated at the same time. The impulses are transmitted in a digital, on/off manner to the electrodes and are fired off rapidly enough so that the wearer perceives the sound to be continuous speech, much as the consecutive frames of motion picture film are projected quickly enough to give the impression of uninterrupted motion. SPEAK was developed before the days of miniature, high-speed digital processors, so it was designed to take samples of those frequency bands and present them to the electrodes. The more recent CIS system, which has become the most widely used auditory implant operating system, has similarities to SPEAK but can present more useful information to the user by availing itself of the capability of modern high-speed electronics.

One of the benefits of CIS that Pierschalla most valued was the ability to hear music much more clearly than when he first received his implant. "With the first analog processor, some music sounded good and some didn't. Some of the stuff I used to really like to listen to like rock-n-roll didn't sound good anymore, so at first I avoided listening to music," he said. Then he realized that he was able to enjoy different kinds of music. Vocals, reed instruments like the saxophone, and the bass guitar sounded good to him, so he started collecting small-combo jazz recordings. The advent of CIS made the whole range of his music accessible to him once again. "In the past, the singing voice didn't sound very good. Even if I knew the words, they wouldn't always sound intelligible. Now if a person is singing clearly enough, like Billie Holiday, I can hear recordings I never heard before and get 70 to 90 percent of the words right." And that, he concluded, "is almost as good as just being able to hear again."

⚊⚊✦⚊⚊

PIERSCHALLA'S KNOWLEDGE AND SUCCESS was widely known within the cochlear implant community. As a result, in 1995 Med-El, a cochlear implant manufacturer based in Innsbruck, Austria, asked him to establish its North American office in Boston, as an organization of one. He accepted, and shortly thereafter he relocated the company's office to Research Triangle Park, North Carolina. He then worked on obtaining FDA approval for sale of the Med-El implant in the United States. That goal accomplished, he turned his attention to developing documentation, such as manuals and patient brochures, creating and traveling with a trade show exhibit, and counseling potential and newly implanted patients.

Then, four years after going to work for Med-El, and twenty-four years after he experienced the first symptoms of Cogan's syndrome, the disease returned with a vengeance. "When he found he could no longer function properly in the job he felt he should be doing, he took a leave of absence," said his father Lawrence, who still lived in Wausau. "He called and asked me to come out and take him back home. He came back to live with me and he remained here until the time of his death."

Even as his health was deteriorating, Pierschalla gave of himself freely to advance cochlear implant technology. There were times he had to use two canes to get on an airplane to visit laboratories where physicians and engineers wanted him to test a new processor or piece of processing software, and all he asked was that they pay his travel expenses. "The last two or three years, he was all over the country flying here and there," said his father. "He never hesitated, regardless of what was going on or how he felt. If he could walk and get on the airplane, he was gone."

At its 2000 Neural Prosthesis Workshop, the NIH Neural Prosthesis Program honored Pierschalla with a plaque that reads, "To Michael Pierschalla, in grateful recognition of enormous personal contributions to the improvement of cochlear implants."

On June 25, 2002, Michael Pierschalla, who had touched so many lives, suffered a massive seizure ending in heart failure. He was 46 years old.

An obituary in the spring 2003 newsletter of RIT's School of American Crafts, read in part, "Michael achieved excellence in several arenas of endeavor. He left great legacies in science and art, and perhaps most admirably, in the hearts of so many. Michael used the suffering in his life to continually deepen his compassion and humility. He lived an examined life, and strove consciously to grow in integrity and humor. The outpouring of love and respect by those who knew him is testament to his high level of personal and interpersonal attainment." In 2003, a group of Pierschalla's colleagues established a scholarship in his name at the Penland School of Crafts, in Penland, North Carolina.

2

The Body Electric

In the mid-1700s, electricity was a novelty, its awesome power having been harnessed in a crude way a century earlier—when the static electricity generator was invented—with little improvement in the interim. And without batteries, only small amounts of electricity could be stored for short periods of time in a Leyden jar. One of the primary uses of electricity was to entertain the upper crust during parlor games at which people would be literally shocked or illuminated. Yet some scientists were also at work trying to understand and harness this strange force for the good of mankind. Among them was Benjamin Franklin, who was respected worldwide as a scientist as well as a founder and representative of an upstart nation.

In fact, Franklin used a Leyden jar in his famous kite experiment in 1752, when he proved that lightning is electric by flying a silk kite tethered by a wet string through a cloud. A metal key was attached to the string, which led to a Leyden jar in which Franklin successfully bottled an electric charge. This led to his invention of the lightning rod, the purpose of which was to prevent fires by attracting lightning away from buildings and sending the charge into the ground. Some fundamentalists of the day were quite incensed at this invention. They thought that if God wanted to unleash fearsome lightning to burn down a building, who were mere mortals to interfere with his design?

Franklin was also one of the first scientists to try to use electricity to return movement to individuals who were paralyzed. He even thought he might be able to cure paralysis by applying electricity to the skin, but he

learned that while he was able to induce uncoordinated movements, a cure
or even coordinated control was out of the question. Franklin may well have
contemplated the idea of creating functional movement in paralyzed individuals, but one of the great limitations he confronted was that his static
electricity–generating machine and the Leyden jar he used in his experiments were not portable. Though he gave up his search for a cure when he
realized that the shocks had no long-term beneficial effects, his interest in
electricity and the study of the relationship between electricity and living
organisms continued to thrive.

Now, hundreds of years later, we take electricity for granted, except
when it isn't there, as when a storm knocks out power lines. It is then—when
we start to wonder if the food in the refrigerator will spoil, and how we will
make it through the day without lights, air conditioning, elevators, the ability
to cook on electric ranges, and without traffic signals—that we begin to
appreciate how electricity has become an integral part of our lives. But it is,
in fact, much more essential to our being than we realize, for our bodies are
electrochemical machines.

The acts of settling into a chair, opening this book, and reading these
words, require the firing of billions of nerve cells, called neurons, in your
brain and throughout your body. At this very moment, as you see, absorb,
and think about what you are reading, cells within various parts of your brain
are firing off tiny jolts of electricity that send signals to neighboring neurons
that together enable you to see and comprehend the symbols on the page
and to contemplate their meaning. Visually processing the words is accomplished primarily by the neurons in your visual cortex in the back of your
brain. As you prepare to turn a page, another set of nerve cells in your motor
cortex, located in the middle of your head, send off action potentials, which
are transmitted along nerves running down your spine to your arm and
hand, where the electricity causes the muscles to contract. While this massive amount of electrical activity is going on within your body, without so
much as a conscious thought, you turn the page.

The process that enables you to do all of this involves chemical responses as well. When an electrical signal reaches the end of a neuron, it
triggers a reaction that releases chemicals, called neurotransmitters, that

travel across the tiny gaps between the mass of neurons that make up your brain. After leaping across these gaps, or synapses, the neurotransmitters attach to nearby neurons and instruct them to activate their own electrical signals, which in turn zip through those neurons to the next set of synapses. In the process, each neuron performs its assigned task of altering the signal until in the end, the aggregate of messages become perception, thought, and movement. Neurons that are connected to sensory receptors, such as your eyes and ears, and send signals to the brain are called afferent neurons. Neurons that work in the opposite direction, sending signals from the brain to muscles or other organs in the body, are called efferent neurons.

The complex of nerve cells in your brain and spinal cord comprise the central nervous system, while those that branch out into other parts of the body to innervate muscles and sensory receptors constitute the peripheral nervous system. The autonomic nervous system, which is essentially a subset of the peripheral nervous system, handles the automatic functions of the body such as digestion, breathing, and blood circulation.

Though each neuron is an individual cell, and we tend to think of cells as being very small, a single neuron can extend over 3 feet in length as it runs to or from the brain through the spinal cord. Neurons also take on a wide variety of shapes. Yet regardless of their size and shape—short and squat, or long and thin—neurons are similar in their makeup to trees; they have branches that reach out from the main trunk, which is the cell body, to other neurons and sensory receptors. One end of the neuron that extends away from the cell body contains one or more little bushy protrusions that pick up the neurotransmitters given off by neighboring cells. These protrusions are known as dendrites, a name derived from *dendron*, the Greek word for tree. Every neuron in your cerebral cortex (the gray matter responsible for high-level functioning that is the outer layer of the brain) contains five thousand to ten thousand dendrites, each about 1 micron in size. Under a light microscope, they look like thorns on a rose bush.

Running in the other direction from the cell body are axons, which carry information from the cell and transmit it to the dendrites of other nearby cells by sending neurotransmitters across the synapses. Though axons do not come in direct contact with dendrites, they are usually no more than several billionths of an inch from them. Most neurons have synaptic connections with thousands of other neurons and receive signals from many of them before firing a jolt of electricity down a dendrite, through the cell body, and along its axon to the next set of adjacent neurons. The spikes of electrical current speed through a nerve cell at about 328 feet per second, and last for one- to two-thousandths of a second. After firing, a neuron is set to do so again within the same period of time. The duration and frequency of these signals are essentially the vowels and consonants of the nervous system's language, since it is these two factors, and not the strength of the electrical signal, that carry the information the neuron is passing along in the complex process of life. Considering that there are billions of neurons in the human brain and most have thousands of connections, the complexity of the brain and its processing of sensations, actions, thoughts, and memories is staggering to contemplate. Yet all those neurons in your brain allow you to do just that—contemplate their, and your, existence.

ALL MATTER IS MADE UP OF ATOMS, and all atoms contain three types of particles: electrons, which are negatively charged particles; protons, which are positively charged particles; and neutrons, which are neutral. Since Mother Nature usually likes an equilibrium, peace is kept in the atomic world by maintaining an equivalent number of electrons and protons in each atom. But, much like restless children, electrons can jump from one atom to another, in which case the affected atom becomes a charged ion. If an ion has more electrons than protons, it is a negatively charged ion. If it has fewer electrons than protons it is positively charged. Electricity is born of the fact that dissimilar particles attract each other, while similar particles repel each other. So, following nature's urge to strike a balance, when a negatively charged ion ends up next to a positively charged ion, the excess electrons in the negatively charged ion will jump over to the one that has fewer of them, until a neutral state is achieved.

In living organisms, ions are the charged particles that flow through tissue creating electrical current. And while a similar process takes place regardless of whether electricity flows through a copper wire or through the human body, the manner in which it takes place in living organisms differs from the way it moves through inanimate objects. Specifically, electricity flows as electrons in conducting metals and as ions inside the body. This distinction is important in the field of neural prostheses and has created a world of woes for those who develop electrodes to inject electricity into the body, because they must generally use metal in the electrodes. But where electrodes meet biological tissue the current becomes ionic. And when electrons, ions, and metal meet, corrosion can take place—definitely not a good thing to be going on inside the body. This dilemma is compounded by the fact that the inside of the body is primarily a saltwater environment; sodium and chloride ions are the most prevalent carriers of electricity within the body. And as anybody who lives where salt is spread on the roads during winter knows, there is nothing like salt to induce rust, which is a form of corrosion. And that is only one of the myriad problems neural prosthetic researchers face in their quest to use electricity to return function to a broken body.

The skeletal muscles that create movement are activated electrically in much the same manner as neurons. Their cells are covered by a membrane that maintains a potential across itself with the inside about one hundred millivolts negative with respect to the outside. When an action potential travels through a neuron embedded in a muscle, the neuron releases a chemical transmitter that causes ionic gates to open in the muscle membrane. These gates allow positively charged ions to temporarily flow in, causing an action potential in the muscle itself. This action potential spreads along the muscle, causing it to contract. When enough cells contract, movement takes place. Similarly in the eye, when light strikes the photoreceptor rods and cones, they convert it to analog electrical signals, which are then processed through several layers of cells in the retina until they become digital electrical signals that are sent through the optic nerve to the brain for further processing. In the ear, the hair cells in the cochlea serve the same purpose for the conversion of sound to electricity. In the mouth, the taste buds do the job, while in the nose it is the olfactory receptors, and in the skin, pressure, heat, and cold stimulate the tactile receptors to begin their electrical firing patterns.

In the final analysis, the flow of electricity in the body quite literally accounts for life itself. Without it, the human body would be like a car with a dead battery. Nothing would function. And it is precisely because the body is so dependent on electricity to function that physicians, scientists, and engineers can now use modern electronic technology to inject electricity into the body to bypass circuitry that has been damaged by injury or disease.

Unfortunately for those in whom neurons become damaged or diseased, nerve cells as a rule do not divide, multiply, and make functional connections after birth. So, unlike skin cells, for example, which multiply rapidly to heal a cut, scrape, or burn, neuronal damage does not heal. If a nerve cell is damaged, it remains damaged, and that is where the field of neural prostheses comes in, with its goal of using modern technology to create implants that are safe, biocompatible, and long lasting to bypass or replace the damaged neurons. Though it is true that scientists are also striving to develop biological methods of nerve regeneration using stem cells and other biological means, it appears likely that neural prostheses will be available—and in some cases, such as the cochlear implant, already are available—before biological repairs become a reality. At the very least, neural prostheses can fill a gap for those now afflicted, before the more "natural" healing processes of the body can be forced to extend to the nervous system.

As with all science, one innovation builds on another. When it comes to neural implants, the ability to create these minuscule miracles of modern technology is built on the foundation of thousands of years of wonder, probing, research, discovery, and invention on the part of many, most of whom could not have imagined the final destination of the voyage of creation they launched and participated in.

3

Of Frogs' Legs
and Transistors

The realization that some mighty power, later to be named electricity, can have a dramatic effect on the human body, as when lightning strikes, goes back to the earliest humans. And though the first reasoning beings knew they could be felled by the wrath of the gods, they had no idea electricity was doing the dirty work, nor for that matter, what electricity was.

The fact that friction on certain materials causes static electricity and that lodestone (the mineral magnetite) attracts certain metals was also undoubtedly observed well before the beginning of recorded history. The first known attempts to explain these magnetic and electrostatic phenomena date back to around 600 BC, when the Greek philosopher and scientist Thales of Miletus suggested that lodestone had a soul and therefore attracted iron. Several centuries later, the Roman poet and philosopher Lucretius, who lived from 99 BC to 55 BC, hypothesized in *On the Nature of Things* that lodestone created a vacuum between itself and iron, thereby drawing the iron to it. Then the physician Galen from Pergamon, who lived from 130 to 200 AD, took a crack at explaining magnetism by claiming that the atoms of iron and those of lodestone were similarly shaped and therefore locked together when put in proximity to each other, much like strips of modern Velcro.

Despite these vain attempts to explain this strange force, the medicinal powers of electricity were realized at least as far back as 50 AD, when the

Roman emperor Claudius's physician Scribonius Largus realized that standing on an electric torpedo fish could ameliorate the pain of certain ailments. For gout, Largus prescribed that "a live black torpedo should, when the pain begins, be placed under the feet. The patient must stand on a moist shore washed by the sea and he should stay like this until his whole foot and leg up to the knee is numb."

More than seventeen centuries passed before scientists began to develop a true scientific understanding of electricity and magnetism. It was left to Cambridge University–educated William Gilbert, the personal physician to Queen Elizabeth I and president of the Royal College of Physicians, to scientifically define the magnetic and electrical phenomena in his book, *De Magnete* (On the Magnet), published in 1600. Galileo was greatly impressed with Gilbert's book, which proposed that the polarity of a magnet was analogous to the poles of the earth. Gilbert's work also seemed to take a poke at the existing wisdom of the time, that the earth was the center of the universe, and it may have helped Galileo develop his idea that the earth revolves around the sun.

Though known mostly for his work on magnetism, Gilbert clearly distinguished magnetism from electricity, and he is the one who gave electricity its name. He did so by borrowing from the Greek word *elektor*, meaning "beaming sun." The related Greek word *elektron* refers to amber, a bright fossilized resin that was widely used in ancient times in jewelry. And because amber is easily electrified by friction, it clearly demonstrated the power of static electricity during those early days of discovery. In fact, until Gilbert's book was published, static electricity was called "the amber effect." Amber was also thought to have substantial medicinal benefits. By applying the rigorous principles of modern scientific research, which were not in common use at the time, Gilbert found numerous materials in addition to amber that had electrostatic properties. He also observed that, unlike a magnet, an electrically charged object had no poles and that while magnetic force can pass through many materials, electricity cannot pass through an insulator, such as paper.

The first electric generator, built by Otto von Guericke, was designed to produce static electricity. A man of many interests, Guericke served in various civic capacities, including burgomaster of Magdeburg, Germany, for thirty-five years. He was a mathematician, an engineer, an astronomer, and an inveterate inventor. A proponent of the Copernican theory that the earth moves around the sun, Guericke studied the emptiness of space and examined the idea that magnetism controlled the movements of the planets. To do so, he invented a suction pump in 1647 and used it to create a vacuum. In one famous experiment, he drew a vacuum inside two copper bowls. Though only the external air pressure held them together, two teams of eight horses could not pull them apart. He also demonstrated that both magnetism and electricity functioned within a vacuum, which was exceedingly important for future developments in electronics, like the vacuum tube.

He built the first electric generator in 1663 by placing a sulfur ball inside a glass globe. As a shaft passing through the ball was rotated, a pad rubbing against it produced static electricity. Over the years, the balls became larger and mechanisms were developed to spin them faster. In 1672, Guericke's machine produced enough electricity to make the sulfur ball glow, thereby creating the first electroluminescence. This device remained the conventional method for generating electricity for over a century.

The problem with Guericke's machine was that electricity from it was available only while it was being spun. There was no way of storing the electricity. Working on this problem at the University of Leyden in the Netherlands was physicist and mathematician Pieter van Musschenbroek, who, in 1745, developed a device consisting of a glass jar partially filled with water and containing a wire that protruded through a cork sealing the jar. When Guericke's generator electrified the wire, an electric charge was stored in the jar, which came to be known as the Leyden jar. The fact that a jar was used as the storage medium was no accident, for at the time electricity was considered a fluid, and what better way to store fluid than in a jar? This precursor of the modern capacitor made experimentation with electricity far more practical and essentially opened the door to the use of electricity for fun, illicit profit, and legitimate medical applications as well.

The ability to generate and store significant electrical charges also presented a significant danger. Musschenbroek told of testing a charged Leyden jar by touching the wire coming from the jar with his right hand: "Suddenly my right hand was struck with such force that my body shook as if hit by lightning. Generally the blow does not break the glass, no matter how thin it is, nor does it knock the hand away; but the arm and the entire body are affected so terribly I can't describe it. I thought I was done for."

—⸺⸺—

DURING THE EIGHTEENTH CENTURY, electricity remained a novelty and a thing of wonder. It was, in modern-day vernacular, "the next big thing," a seemingly magical force that generated much glee at parlor games of the well-to-do, and it was also considered a cure for many of humankind's illnesses. Charlatans and hucksters as well as legitimate scientists and physicians were experimenting with electricity and expanding its potential uses, some for legitimate medical applications, others for profit. For merriment, people electrified all sorts of objects, even other people. One favorite demonstration was to suspend a youth by ropes, apply an electric charge to his body, and excitedly watch as he attracted bits of metal. Sometimes children were partially covered with resin and then electrified, which caused them to give off a saintly glow. Electricity also served as the whoopee cushions of the day; hostesses would sometimes give their guests an unexpected charge and then make up for it with a peck on the cheek.

On the medical front, electricity was applied to aching teeth and was credited with the ability to cure headaches, sciatica, gout, rheumatism, and menstrual discomfort, among other ailments. It was also used to magically administer drugs, such as purgatives. The patient held the potion in one hand while being given an electric shock. The concoction was said to then pass directly into the patient's body without being diluted by oral administration. Despite such quackery, a number of serious scientists were exploring the use of electricity for medical purposes. One was Jean-Antoine Nollet (1700–1770), a French clergyman and physicist. In 1748, he attempted to use electrical shock to cure three paralyzed individuals by administering electricity in a variety of ways. His attempts were likely based on previous observations that application of electricity caused muscles to twitch involuntarily. Though he observed no long-term effects, he was encouraged enough by his experiments to write, "Electricity perseveringly employed and skillfully administered can be a useful remedy for the paralysis located in nerves or muscles." Since Nollet and Benjamin Franklin lived and worked during the same period, it is likely that they had heard of, and perhaps learned from, each other's work.

Another scientist who focused on electricity and paralysis was Luigi Galvani, a physician and professor of anatomy and obstetrics at the University of Bologna, Italy. In the late 1700s, Galvani used an electrostatic machine and a Leyden jar as his tools to conduct experiments, primarily by stimulating the nerves and muscles of dissected frogs. During one such experiment, on January 26, 1781, an electrostatic machine happened to be on the same table as the lower extremities and exposed nerves of a frog. An assistant of Galvani's noticed that when a metal scalpel came in contact with the nerves, the frog's legs went into convulsions. Trying to understand this serendipitous occurrence, Galvani attempted to reproduce the twitching by touching the exposed nerves with nonconducting materials such as glass, both when the generator was sparking and when it was not running. He found that the muscles twitched only when the generator was sparking and the nerves were touched with a conducting material, like the metal scalpel. To his mind, this phenomenon supported his earlier-held theory that "a highly subtle fluid exists in the nerves." He rightly thought that the body generated its own electricity. He called this fluid animal electricity and reasoned that in the case of the frog's legs, it was "excited into motion by the impact, vibration, and impulse of the spark."

This led Galvani to wonder whether electricity in the outside atmosphere during a thunderstorm would have the same effect on frogs' legs. To find out, he attached several of them to metal wires and hung them outside during an electrical storm. The result was that, just as in his laboratory, the leg muscles twitched. In another experiment, he attempted the same thing on a clear day. In this case, Galvani hung his prepared frogs from iron hooks over a railing. When the hooks came in contact with the iron of the railing, the frogs' legs jumped, even though there was no electricity in the air. Galvani hypothesized that the muscles were acting like an organic Leyden jar, storing whatever electricity was in the air, which was discharged when the hooks touched the railing. To test this theory further, he brought his experiments back inside the laboratory, and this time he held a frog by the spinal cord, touching its feet to a metal box. With his other hand, he brought another piece of metal into contact with the box. The frog jumped. To Galvani, this gave further credence to the Leyden jar effect, that muscles and nerves attract and retain animal electricity.

Alessandro Volta, a professor of physics at the Royal School in Como, Italy, and a contemporary of Galvani's, differed. It was his hypothesis that the electricity firing the frog's legs was generated by two dissimilar metals coming in contact with a moist substance and that the frog's legs provided the salt needed for the electrical reaction. He believed that rather than electricity residing in the nerves and muscles, as Galvani felt he had proven, the electricity was externally generated. The debate between the two intellectual giants became a cause célèbre among educated Europeans, although the two remained respectful and gentlemanly toward each other.

In the process of developing experiments to prove his hypothesis, and disprove Galvani, Volta invented the battery, named the voltaic pile in his honor. The first voltaic pile was a stack of alternating silver and zinc disks, interspersed with soaked pieces of pasteboard. Volta found that other metals such as copper, brass, and tin, would work as well. When the ends of these piles were touched simultaneously, a slight shock was felt. Once he was able to use his battery to create a steady flow of electricity on demand, he went about testing the effects of that electricity on his own body. He found that "the current of the electric fluid . . . excites not only contractions and spasms in the muscles, convulsions more or less violent in the limbs through which it passes in its course; but it irritates also the organs of taste, sight, hearing, and feeling, properly so called, and produces in them sensations peculiar to each." In one particularly dramatic experiment, he placed two charged probes into his ears: "At the moment the circuit was completed I received a shock in the head, and a few moments later . . . I began to be conscious of a sound, or rather a noise, in my ears that I cannot define clearly; it was a kind of jerky crackling or bubbling, as though some paste or tenacious matter was boiling." Wisely, he wrote, "I did not repeat this experiment several times."

We now know that there was truth in the theories of both Galvani and Volta; externally generated electricity can stimulate muscles, and the nervous system itself also generates its own electrical signals. But at the time, neither of the scientists was willing to give an inch. As Galvani wrote of the controversy, "He wants this to be the same electricity that is common to all bodies; I want it to be special and specific to the animal. He locates the cause of the imbalance in the contrivances used, particularly in the dissimilarity of metals; I, in the animal machine. He determines such a cause to be accidental and extrinsic; I, to be natural and internal. He, in short, attributes everything to metals, nothing to the animal; I everything to the latter, nothing to the former." Despite their differences, however, to honor his intellectual rival, Volta termed the effect externally generated electricity had on nerves and muscles "Galvanic stimulation."

A primary goal of Galvani's experiments was to learn how to treat paralysis. One problem he faced was determining what precisely goes wrong with nerves when paralysis occurs. Of this quandary he wrote, "As to the treatment of paralysis, I see a problem full of difficulty and danger; for it is not easy to diagnose whether the disease originates from damage to the structure of the nerves or cerebrum, or from the presence of a nonconducting substance in the inner parts of the nerve or in other bodily parts wherein we believe an electric circuit is completed. If the first, artificial electricity, in whatever way it is employed, can be of little use, and perhaps can even be injurious; if it is the other, it seems possible some benefits can be derived from dissipating the nonconducting substance or by increasing the strength of animal electricity." He did not know that paralysis is caused by "damage to the structure." And contrary to his conclusion, it is, in fact, the use of "artificial electricity" introduced by neural prostheses that is helping to overcome paralysis.

Playing on Galvani's science, charlatans were also at work, several of whom independently developed "galvanic spectacles" that had dissimilar pieces of metal in the nosepiece and were said to deliver a salubrious jolt to the optic nerve. One slightly less optimistic developer of such spectacles claimed only to clear the nasal passages.

Such nonsense aside, Galvani and Volta greatly furthered the understanding of electricity and its effects on the human body, and they made significant steps toward the creation of implantable devices that can interact with the nervous system to overcome damage and enhance sensory and motor capabilities. The work of these men also fired the imagination of Mary Shelley, the wife of the poet Percy Shelley and the author of *Frankenstein*, which was first published in 1816. In an introduction to her classic horror story in which she describes the evolution of the idea for her famous story, she writes, "Perhaps a corpse would be reanimated; galvanism had given token of such things; perhaps the component parts of a creature might be manufactured, brought together, and endued with vital warmth." And in the novel itself she wrote, "It was on a dreary night of November that I beheld the accomplishment of my toils. With an anxiety that almost amounted to agony, I collected the instruments of life around me, that I might infuse a spark of being into the lifeless thing that lay at my feet." The reanimation of a corpse through galvanism and infusing a "spark of being" were not ideas unique to Shelley, as some had actually attempted to bring back the dead using electricity during the early nineteenth century.

On the scientific front, the march toward a fuller understanding of electricity and how to harness its power continued when in 1820 Hans Christian Oersted, a Danish physicist, demonstrated the interaction of electricity and magnetism by showing that an electrical current in a wire will cause a compass needle to shift to a right angle to the wire. This gave birth to the field of electromagnetism and opened the door for Michael Faraday, a largely self-taught English scientist, to build the first electric motor in 1821. Ten years later, Faraday discovered electromagnetic induction—the basis of the electric generator and the transformer, which allows an electric current to be induced in a conductor in a magnetic field. This principle is used in modern neural prostheses to pass power and information to and from devices implanted in the human body.

Faraday's discoveries and inventions as a whole made it possible for electricity to go from a novelty to a force that would permanently alter the direction of society. Faraday also coined the words *ion*, *electrode*, *electrolyte*, *anode*, and *cathode*.

The French physician Guillaume Benjamin Amand Duchenne brought Faraday's induction principle to the medical field by using it to generate pulsating electrical current, which he fed to electrodes placed on the surface of the skin. He was then able to precisely measure the amount of current needed to activate muscles. To accomplish this, he used a portable electrical machine in a box that he designed, which he carried to his patients' bedsides. While others before Duchenne had applied electricity to the skin to induce muscle contractions, he was the first to do so in a scientifically controlled manner. By observing the stimulated muscles, he was able to learn a great deal about muscular anatomy. As a result of this seminal work, Duchenne is credited with the development of electrodiagnosis and electrotherapy and with laying the foundation for the specialty of neurology. Duchenne was also an excellent photographer who photographed the expressions of some of his subjects as he applied electrodes to their facial muscles causing a variety of contorted expressions. Though much of his work was published in the 1860s, and he died in 1875, researchers currently developing functional electrical stimulation systems to give motion to the paralyzed still find the results of Duchenne's muscular studies useful.

While Duchenne was busy mapping the human musculature, a fellow physician, Gustav Theodor Fritsch, began mapping the human brain. While treating soldiers wounded during the Prusso-Danish War of 1864, the German physician noticed that when he happened to touch the cerebral cortex of the brain while dressing head wounds, some of the patients' muscles would twitch on the opposite side of the body. When he returned to his laboratory, Fritsch investigated this phenomenon with his colleague Julius Eduard Hitzig by using induction coil technology to stimulate various parts of the brains of animals. They found that the brain contains a map of the entire body, with different areas of the body represented in different parts of the motor cortex of the brain. It was through Fritsch and Hitzig's work, as well as that of other scientists studying the brain at the time, that a picture began to emerge of which parts of the brain process specific inputs and which are responsible for specific outputs.

In 1874, a physician named Roberts Bartholow took this work a major step forward by inserting an electrode into the brain of a woman whose skull was eaten away by cancer of the scalp. By feeding current through the electrode and relocating it to various parts of the woman's brain, Bartholow was able to induce muscular contractions in various parts of her body. Writing of the reaction when the electrode was placed in the woman's right cerebral cortex, Bartholow stated that "her countenance exhibited great distress, and she began to cry. Very soon the left hand was extended as if in the act of taking hold of some object in front of her; the arm presently was agitated with clonic spasms; her eyes became fixed, with pupils widely dilated; lips were blue, and she frothed at the mouth; her breathing became stertorous; she lost consciousness, and was violently convulsed on the left side."

Not surprisingly, a number of Bartholow's colleagues thought he had gone too far with these experiments, and the American Medical Association censured him for his actions. Cruel as they may have been, they did nonetheless establish an unquestionable link between electricity and the functioning of the human brain.

While initial studies involved placing electrodes on or in the cortex—the outer layer of the brain—it wasn't long before researchers began to explore the workings of the inner portions of the brain by sinking electrodes deeper into the brains of animals to test for responses to stimulation. At first, the animals had to be held still for these experiments. Then, during the early 1900s, Walter Rudolf Hess, a Swiss physician, developed a method of attaching electrodes to animals' skulls that allowed for stimulation of deep regions of the brain while the animals were unfettered. By doing so, Hess found he could elicit not only sensory responses but behavioral responses as well, such as sexual stimulation, urination, grooming, and feeding.

Hess won the 1949 Nobel Prize in physiology for his studies of the workings of the diencephalon, the portion of the brain made up of the thalamus and the hypothalamus. By stimulating the brains of cats, he found that the thalamus processes sensory information, while the hypothalamus plays a key role in a number of automatic physical functions, such as breathing, digestion, sneezing, and the regulation of blood pressure, and is also involved in emotions and biological drives, including fear, anger, and sleep. These findings helped bridge the gap between physiology and psychiatry.

Following on the heels of Hess's work was psychologist James Olds, famous for his 1950s experiments that involved stimulating the brains of rats. Olds was familiar with the work of others who had implanted the brains of animals to cause them to try to avoid being stimulated by the electrodes. He wondered if he could accomplish the reverse, that is, stimu-late pleasure zones in the brain to get the animal to behave in a way that would bring on stimulation. As a postdoctoral fellow at McGill University in 1953, he began his study of the reticular system—a part of the midbrain near the diencephalon that controls wakefulness and sleep.

To conduct his studies, Olds rigged up a rudimentary electrode system that he implanted in the brain of a rat, hoping to stimulate the reticular system. During implant surgery, the electrode fortuitously missed its in-tended target and landed in the rhinencephalon. When Olds began stimulat-ing the rat's brain, the rat was standing in a corner of a box. Olds noticed that subsequently the rat kept returning to the same corner of the box, apparently hoping to be stimulated again. When Olds changed the corner in which the rat received the stimulation it returned to the new corner with increasing frequency. The stimulation was seemingly so pleasant that the rat chose it above food, even after not eating for twenty-four hours.

Taking the experiment to the next level, Olds installed a lever that al-lowed the rat to stimulate itself, which it did about once every five seconds. When the current was turned off so that pushing the lever had no effect, the animal just lay down and went to sleep. When the current was turned on again, "and the animal was given one shock as an *hors d'oeuvre* it would begin stimulating its brain again," wrote Olds.

The discovery made by this work came to be known as the "brain re-ward" system. Olds went on to study and map the reward systems of various other areas of the brain. Some of the highest reward values were achieved by stimulating parts of the hypothalamus. When this area of the brain was implanted, the rat would hit the lever as many as two thousand times per hour. Little wonder it needed some sleep when the current was turned off. Continuing along these lines, Olds investigated what other kinds of actions could be induced by stimulating various parts of the brain. He was, for example, able to stimulate the urge to eat and drink. He also used electrodes implanted in the brain to study learning and memory.

Two pioneers who took this work to the human level were neurosur-geons Otto Foerster and Wilder Penfield. Working separately, Foerster, a German, and Penfield, a Canadian, conducted their experiments on patients whose brains were exposed because they were undergoing brain surgery for diseases, such as cancer, or who were having surgery to remove portions of the brain causing severe epileptic seizures. These patients had to be kept awake so that they could provide feedback to the surgeons to ensure they were not damaging or destroying healthy tissue. This was possible because, though the brain is the organ that receives and processes pain information from throughout the body, it has no pain receptors itself. So probing the brain in a conscious patient causes no discomfort, unless the pain-processing center for another part of the body is impinged upon. Having the brain exposed out of necessity presented a perfect opportunity to conduct ancillary experimentation on the patients, who could provide feedback as to what they were experiencing—something animals could, of course, not do.

Working in the 1920s, Foerster is credited with being the first person to electrically stimulate the occipital lobe, where vision is processed. By doing so, he found that a patient saw a spot of light, called a phosphene, which appeared to be in different positions in the visual field depending upon where the electrode was placed. Penfield, who conducted many of his experiments in the 1950s, induced twitching in his human patients, who told him that they were powerless to control the actions initiated by the electrodes. Depending upon where their brains were stimulated, his patients also reported experiencing pain, pleasure, anxiety, flashbacks, sounds, or flashes of light—essentially the entire range of human experiences. It's all there, either stored and activated or created by electrical stimulation. This work added significantly to the understanding of which parts of the human brain are responsible for various actions, perceptions, and emotions, all of which helped to pave the way for later scientists to know where to place electrodes on and in the brain to replace damaged portions of the nervous system.

Psychiatrist Robert Heath of Tulane University took human implantation into the behavioral realm, as Olds had done with rats, and in doing so created considerable controversy. Among other things, he attempted to cure homosexuality—which during the mid-twentieth century was considered a malady—by implantation. In one of his experiments Heath reportedly implanted an electrode in the pleasure center of a gay man, who, like Olds's rat, was able to self-stimulate, and reportedly did so at a rate of 1,500 times within three hours. In a 1972 article in the *Journal of Behavior Therapy and Experimental Psychiatry*, Heath stated that "during these sessions, B-19 [the patient's identity number] stimulated himself to a point that he was experiencing an almost overwhelming euphoria and elation, and had to be disconnected, despite his vigorous protests."

In another case, Heath implanted a patient who was frequently unmanageable and had to be restrained in a straightjacket. By stimulating his cerebellum through implanted electrodes, the individual's violent outbursts were stemmed, but only as long as the current was fed to the electrodes, and no lasting benefits were realized.

During the early 1960s, neurophysiologist W. Grey Walter made the surprising discovery that there is an increase in neural activity in the brain signaling that one intends to move a limb, even before the individual is consciously aware of the planned movement. This advance firing, called the "readiness potential," was discovered when electrodes were temporarily placed in the brains of individuals who were having brain surgery for epilepsy or some other disease, as Foerster and Penfield had done. Walter fed the output of the patients' brains to a slide projector and found that their readiness potentials activated the projector even before the muscles of their arms contracted to move toward the switch.

Meanwhile, Yale University neurophysiologist José Delgado put his life on the line to prove the ability to control animal behavior by means of electrical stimulation of the brain. In his book *Physical Control of the Mind*, Delgado describes implanting radio-controlled electrodes in the caudate nucleus of aggressive monkeys. When the transmitter was turned on, causing electrical stimulation of the brain, the monkeys became docile. He then placed a lever to control the electrodes in a cage populated by an aggressive male monkey and several submissive monkeys. It didn't take long for a female monkey in the cage to discover that the dominant monkey's aggression was inhibited when she pressed the lever. She even went so far as to stare right at him while she did so, an act of defiance that otherwise would have brought on immediate retaliation.

Demonstrating fearlessness himself, Delgado, a native of Spain, got into a bull ring with an aggressive bull whose brain had been implanted with radio-controlled electrodes. After testing the apparatus on the bull from a distance, Delgado decided to play matador for a day. He positioned himself in front of the bull and taunted it, as would any matador. But Delgado had a secret weapon. As the bull charged, he used a remote control to turn on the electrodes in the bull's brain, causing the bull to come to an abrupt halt and turn instead from side to side.

Though early behaviorally oriented implantation of the brain generated considerable controversy, and has since been abandoned, the work of these pioneering brain researchers paved the way for those who followed.

The first two medical implants to be accepted for widespread clinical use were the heart pacemaker and the phrenic nerve stimulator, both developed during the 1950s. Technically speaking, the pacemaker is not considered a full-fledged member of the neural prosthetic family, since its major function is to control the heart muscle rather than a nerve, even though it can be argued that the pacemaker stimulates neural fibers in the heart. Nonetheless, experts in the field consider the phrenic nerve stimulator—which activates the nerve providing electrical impulses to the thoracic diaphragm, allowing patients who cannot breathe on their own to be taken off ventilators—the first neural prosthetic device to be clinically implanted in patients.

Early versions of the pacemaker and the phrenic nerve stimulator were connected to their external controllers by wires that ran through the patients' skin. These percutaneous connections posed a risk of infection. The heart pacemaker also used vacuum tubes and large batteries that had to be wheeled around on a cart with the patient. During the late 1950s, William Glenn, chief of cardiovascular surgery at Yale, and his colleagues, developed an implantable phrenic nerve stimulator that, based on Faraday's findings over one hundred years earlier, received power and information by means of radio waves sent through the skin. Glenn and his colleagues also played a major role in developing radio frequency–powered pacemakers.

The first implantable auditory prosthesis, a precursor of the modern cochlear implant, was successfully implanted in a patient by two French researchers, Andre Djourno and Charles Eyries, in 1957. The recipient was a totally deaf individual whose diseased cochlea had been surgically removed.

The French physicians permanently implanted an induction coil in the patient's temporalis muscle, which operates the jaw, and an electrode on the auditory nerve. Another coil, placed on the skin over the implanted coil, induced stimulation current in the internal coil, not unlike the methods used with modern cochlear implants. The patient initially said he heard sounds like crickets or a roulette wheel. Over time, he was able to discern speech rhythms, which aided him in lipreading, and eventually he could distinguish some words. Perhaps most importantly, the patient took pleasure in being able to hear sounds of any type.

While the prognosis was optimistic, this experiment did not open the floodgates to implantation. There was substantial concern over the long-term effects of permanent implantation, and the technology did not yet exist to make such a practice feasible on a wide scale. Eventually the French team dropped the auditory implant idea because their electrode leads tended to break; they didn't have adequate biocompatible materials that could withstand the corrosive saltwater environment found inside the body. In essence, their biological concept was ahead of the technology.

Another auditory implant pioneer was Blair Simmons, a neurophysiologist and otolaryngologist at Stanford University, who implanted six electrodes in the cochlea of a deaf individual in 1964. One of his goals was to determine if multiple electrodes would allow the patient to identify a wider range of sounds than could the French researchers' subject. They did, as Simmons's patient was able to make out differences in some speech sounds and could even identify simple tunes. Despite this success, Simmons did not undertake additional human implantation but rather continued animal experiments to determine the effects of long-term stimulation of the auditory nerve.

The first time electrodes were actually placed on the brains of humans in an attempt to return a lost function involved stimulation of the visual cortices of three blind individuals. These experiments were reported in 1962 by two Iowa researchers, J. Button and T. Putnam. Their visual prosthesis was a crude system consisting of four electrodes that passed through the scalp and received signals based on the output of photocells. The implanted volunteers were able to use the photocells to determine the location of a group of light bulbs and to locate the light given off by candles on a birthday cake. The implants proved that stimulation of the visual cortex could induce some sort of visual perception, but controlling the perceptions and making them practical were still major challenges. The electrodes were removed from the subjects after several weeks. Then, in 1967, British physician Giles Brindley and neurosurgeon Walpole Lewin permanently implanted eighty-one electrodes on the visual cortex of a blind nurse, who was subsequently able to see spots of light.

While physicians were making early stabs at neural implantation with whatever equipment they had at the time, other scientists were at work advancing electronic technology by shrinking components and making them faster and more reliable. It is doubtful that any of them had neural prostheses on their minds as a potential application for the technology they were developing, yet the two fields eventually merged.

AS THE TWENTIETH CENTURY DAWNED, there was an explosion in the evolution of electronic technology that facilitated the work of researchers who wished to peer inside the brain and other parts of the nervous system to learn its secrets. Over a period of about fifty years, in what has come to be known as the Electronic Age, the electron was discovered, and the radio, the vacuum tube, the transistor, and the integrated circuit were all invented. And with each of these inventions, the time drew significantly closer in which medical science could seriously contemplate building minuscule devices to permanently place inside the human body to repair damaged tissue as delicate as the retina.

The first flicker of intellectual light leading to the dawn of the Electronic Age flashed in the laboratory of Thomas Edison, when in 1884, while striving to improve the incandescent light bulb he had invented five years earlier, Edison found not only that current flowed through the filament of the bulb, but that it could be made to flow through the vacuum inside the bulb as well and that it flowed in only one direction. The English physicist Joseph John Thomson picked up on this work and built a vacuum tube, similar to the light bulb, to more thoroughly investigate what was called the Edison Effect.

In 1897, he found that the particles flowing through his vacuum tube, which he called "corpuscles," are contained in all matter. Thomson had discovered the electron, for which he received the 1906 Nobel Prize in physics.

Meanwhile, John Ambrose Fleming, another English physicist, invented a two-electrode radio rectifier, which he called the oscillation valve. Patented in 1904, this device is now known as the diode. Wireless transmission was in its infancy at the time, and the diode helped detect radio waves and convert, or demodulate, them into electricity. It is the diode that now makes it possible to power neural prosthetic devices implanted in the body by radio waves transmitted through the skin, precluding the need for wires that pass through the skin or batteries that may need replacement.

Radio waves were first generated and detected in 1894 by Heinrich Hertz, whose transmissions spanned the length of his laboratory. Within a year of Hertz's accomplishment, the Italian scientist Guglielmo Marconi increased that distance to almost 2 miles. In 1896, Marconi received a patent for his system, and in 1897, he founded the Wireless Telegraph and Signal Company, which became Marconi's Wireless Telegraph Company Limited in 1900. The following year, he successfully sent wireless signals across the Atlantic Ocean, a distance of about 2,000 miles, from Poldhu, in Cornwall, England, to St. John's, Newfoundland. For his work, he shared the 1909 Nobel Prize in physics with Karl Braun.

To this point, radio transmissions were used for telegraphy only. Then, in 1906, the American physicist Lee de Forest developed the vacuum tube triode, which he called the audion. It was similar to the diode but had three elements instead of two. The triode, which could amplify as well as rectify radio signals, opened the door to the transmission of radio waves via amplitude modulation (AM), which enabled the transmission and reception of music and the human voice in addition to telegraphic dots and dashes. De Forest, who founded the De Forest Wireless Telegraphy Company in 1902, recast his company as the De Forest Radio Telephone Company in 1907 and the following year transmitted radio signals from the Eiffel Tower that were heard 500 miles away. Then, in 1916, he began a regular radio program, broadcasting music five nights a week from New York City.

The creation of such sophisticated electronic equipment brought not only music and news to the masses. It also gave medical researchers the ability to develop sophisticated electrodes with which to stimulate and record from nerves, and devices such as oscilloscopes, with which they could closely observe the activity of the nervous system. Knowledge of the nervous system's functioning, therefore, grew concurrent with the ability presented by the new electronic technology to more deeply investigate its intricate workings.

The advent of the computer was a huge step toward the development of neural prostheses, although the first electronic computer was a bit too bulky to be of value to the field. Dubbed ENIAC (electronic numerical integrator and calculator), it contained 18,000 vacuum tubes and filled several rooms at the University of Pennsylvania, where it became operational in 1946. Despite its size, it was not as powerful as a modern pocket calculator. Yet the power of the electronic computer was about to increase dramatically, and its size decrease in equal proportion, all of which was made possible by the invention of the transistor in 1947.

The work of the three Bell Laboratories scientists who invented the transistor was predicated on the knowledge that certain crystalline solids have the ability to conduct electricity. Since these crystals do so better than an insulator, such as rubber, but not as well as a conductor, such as copper, they were known as semiconductors. John Bardeen, Walter Brattain, and William Shockley were searching for a basic understanding of how semiconductors work and also had an eye on developing a more reliable replacement for the vacuum tubes, which were ubiquitous throughout the Bell Telephone system, when they struck upon their 1956 Nobel Prize–winning invention.

The Bell Labs physicists focused their work on germanium and silicon semiconductor crystals, and, in fact, the first transistor was made of germanium. The team struggled for some time to get their stubborn piece of germanium to come to life. Then, on December 16, 1947, in what was just one more attempt to induce electrons to act on the crystal, Brattain came up with the idea of pressing two strips of slightly separated gold foil into the germanium. When power was applied, there was a power gain. Amplification had taken place, and the transistor was born.

A few days later, Shockley sent a hugely understated note to five members of Bell Labs' top management inviting them to his laboratory "to observe some effects . . . I hope you can break away and come." For the demonstration, the germanium amplifier was inserted into a communications circuit in place of a vacuum tube, and as Brattain wrote, "a distinct gain in speech level could be heard . . . with no noticeable change in quality."

As with all newborns, the transistor entered the world without a name until several months later, when Brattain was discussing potential names with John Pierce, a Bell Labs colleague. "I presented the problem to Pierce," recalled Brattain. "After some thought, Pierce mentioned the important parameter of a vacuum tube—transconductance—then a moment later its electrical dual-transresistance. Then he said 'transistor,' and I said, 'Pierce, that is it!' "

With a proper name, the transistor was ready to be introduced to the world, so a press conference was held in New York City on June 26, 1948, at which the new device was demonstrated. Except for a few in the technical community, the news was met with a yawn. Even the venerable *New York Times* had nothing more than a small blurb in its "News of Radio" column, which carried the headline "Two New Shows on CBS Will Replace 'Radio Theatre' During the Summer."

The mention read in part, "A device called a transistor, which has several applications in radio where a vacuum tube ordinarily is employed, was demonstrated for the first time yesterday at Bell Telephone Laboratories . . . In the shape of a small metal cylinder about a half-inch long, the transistor contains no vacuum, grid, plate or glass envelope to keep the air away. Its action is instantaneous, there being no warm-up delay since no heat is developed as in a vacuum tube."

At one half-inch long, that first transistor was huge by comparison to current devices, some of which are one hundred times smaller than the diameter of a human hair. And early transistors were produced one at a time, much like individual vacuum tubes, and then soldered to circuit boards along with other components, including resistors, capacitors, and diodes. These circuit boards became large and difficult to assemble and could not take full advantage of the rapid processing ability of the transistor. Then, during the late 1950s, Jack Kilby of Texas Instruments and Robert Noyce of

Fairchild Camera, developed a way to incorporate transistors and other electronic components on a single piece of semiconductor material using a process called photolithography, which is similar to photographic printing. Their invention was dubbed the integrated circuit, which allows millions of transistors and other electronic devices to be packed onto a minuscule chip thousands of times smaller than the first transistor. These modern devices generate almost no heat, are extremely reliable, require considerably less power than vacuum tubes, and because of their extremely small size allow electrons to flow rapidly through them, thereby permitting electronic equipment to operate at higher frequencies.

This same microelectronic circuitry, built on the foundation of thousands of innovations created over hundreds of years by a virtual army of geniuses, has made it possible for contemporary scientists, engineers, and physicians to design and build tiny electrodes, receivers, transmitters, and controllers that can be implanted inside the human body to mimic its own functions. And the fact that electrodes can be made many times smaller than a human hair now gives investigators the ability to record from and stimulate small groups of neurons, rather than just regions of the brain and the peripheral nervous system. Being able to conduct experiments on such a small scale has greatly enhanced the understanding of the effects of artificial stimulation on the nervous system, including the amount of current required to activate nerves without causing tissue damage and how nearby neurons react when specific cells are targeted.

But back in the mid-twentieth century, it was clear that for neural prostheses to be made safe, functional, and durable, much remained to be done. Those goals were left to what could be considered the second coming of neural prosthetics. Following a spate of activity during the 1950s and '60s, there was a period during which the field seemed to lay relatively fallow. This lull can be attributed to a number of factors. For one thing, until 1976 the U.S. Food and Drug Administration did not regulate devices implantable in humans. Prior to that time, if a surgeon wanted to do an implant, he had virtual carte blanche. Additionally, the hippy days of the '60s saw the advent of psychedelic drugs, which brought about a societal backlash to the idea of altering the brain through implants or lobotomies. Neural prosthetic developers also needed new materials that would be more biocompatible and longer lasting than what was available at midcentury. And finally, much less was known about the brain. Once new electronics merged with new materials and increased knowledge, a new group of researchers reinvigorated the field with a fresh sense of excitement. In the words of one of the new flock of neural prosthetic researchers speaking as the twentieth century came to a close, "I think you could make a pretty good argument that we are just now getting the tools available to make clinical impacts."

4

The Grandfather
of Neural Prostheses

A n often-told tale about Giles Brindley might reveal something about the
person referred to as the grandfather of neural prostheses. In 1983, the
inveterate innovator and self-experimenter stood before a scientific audience
and removed his pants. The venue was Las Vegas, Nevada, and the audience
that witnessed this occurrence was the membership of the American Urolog-
ical Association. Brindley was demonstrating, quite graphically, the success
of an injection of phenoxybenzamine, a treatment he had developed for
erectile dysfunction. "The drug is very long acting and I injected it before the
meeting. I wasn't intending to show it," said Brindley, "except the chairman
asked me to, so I asked the audience if they had any objection and the
audience didn't have any objection. I said, 'Well alright,' and I showed it."

Though this has nothing to do directly with neural prostheses, it shows
what a unique and dynamic individual Brindley is. Born on April 30, 1926,
in Woking, England, Brindley single-handedly invented an eighty-one-
electrode prosthesis that in 1967 he and surgeon Walpole Lewin implanted
on the visual cortex of a blind nurse, which succeeded in causing her to see
spots of light. At the same time, he was busily at work on neural prostheses
that were successfully implanted in a number of paralyzed patients to help
them move their arms, stand, walk, and even pedal a bicycle. Brindley is also
the inventor of a bladder and bowel stimulator that allows paralyzed individ-

uals to void more normally and in some cases have erections as well. This neural prosthetic device is commercially available and has been implanted in thousands of individuals worldwide. He is also the founder of the Neurological Prosthesis Unit of Britain's Medical Research Council.

A man of insatiable curiosity, in addition to pioneering the development of numerous neural prosthetic implants, Brindley has conducted many dramatic experiments on his own body. One such experiment was inspired by the fact that birds are known to have color filters in front of their retinas and humans do not. This led Brindley to wonder whether various wavelengths of light would be perceived differently from behind the eye than they are from the front. And what better way to answer that question, he reasoned, than to shine various colors of light onto the back of his own eyeball? Asked if this required removing his eyeball from its socket, he laughingly replied, "Oh, no. You don't need to do that. You only need to turn the eye as far to one side as it will go, and put a light guide into the conjunctival sac—the sac separating the eyelid from the eye, into which tears flow—you can get the light 'round to the back of the eye. No open surgery was required," he said.

"For the first trial, I used an auriscope light, which is a tiny electric light bulb at the end of a rather long stem. I anesthetized the conjunctival sac because it is a bit painful to put an electric light bulb into your conjunctival sac, and then turned the eye right around, put the thing into the sac, and pushed it against the globe, and switched the light on. I could see it immediately, not where it physically was but projected out into space," Brindley said. "To test whether colors looked the same whether the light reached the rods and cones from in front or behind, I had to replace the auriscope bulb by a glass light-guide, and get the light from a prism spectrometer or from a gas discharge tube with color filters to isolate single spectral lines. The answer was that light of any one wavelength looked very nearly the same color, whether it came from behind or in front of the retina."

In that experiment, conducted in 1959, he needed to turn his eye only as far as he could do so voluntarily. In yet another visual experiment conducted the following year, however, he pushed things a bit further. This time, he wanted to learn about the position-sensing ability of the eye. To do so, he enlisted the help of a colleague, Patrick Merton, who, after anesthetizing

Brindley's conjunctival sac, actually seized the eyeball with a forceps. "When he grasped my eye with forceps and pulled it about, I didn't know in which direction he was moving it, and that's the chief finding on passive movement," said Brindley. That is to say, "We find that when the conjunctival sacs are anesthetized, the subjects know neither the amount nor the direction of any deviation of their eyes unless that deviation is being produced by the unhindered action of their own eye muscles."

Years later, Brindley did not remember which eye the experiments were performed on but noted that "I did have a habit of using the left eye as the experimental eye, and so it is quite likely the two had been done on the left, but my eyes at that time were equally good. They are not now. I have glaucoma, and my right eye is a good deal worse, so the one that I experimented on is my better eye now, but at the time they were exactly equal." And there were yet other eye experiments. As Brindley said, "If you are a self experimenter, you are a self experimenter."

Several years earlier, he repeatedly exposed his left eye to very bright light over a two-year period. When asked the purpose of that experiment, he replied briskly, "Science." Following that series of experiments, he experienced what he calls an "after image. A very slight alteration of function of that retina, which lasted for about fifteen years thereafter, but then it recovered. One could hardly call it harm, even while it lasted, but it was obvious to me that the color vision in that eye was different in the region that I had treated in this way. That's the only harmful effect I've ever had."

Asked whether he was concerned that he might inflict serious damage on himself through his self-experiments, Brindley replied with an emphatic, "No! I don't know why, but I wasn't. I was confident that what I was doing was harmless."

———⟨⟨⟨∫∩⟩⟩⟩———

BRINDLEY ATTENDED CAMBRIDGE UNIVERSITY for both undergraduate and medical school and did his clinical work at London Hospital, now called the Royal London Hospital, in East London. But even medical school was not enough to keep his all-encompassing intellect fully occupied. While studying medicine, he also seriously contemplated what makes stars twinkle, to the point that he wrote a paper on the subject entitled "The Scintillation of Stars," which was published in *Nature* magazine in 1950. As an even younger schoolboy, he taught himself Russian. In the process, he found that learning Cyrillic was "trivial. The difficult part is learning the language," he said. He made good use of his knowledge of Russian when shortly after his 1951 graduation from medical school and while serving in research posts in physiology at Cambridge, "I offered myself to the *British Abstract of Medical Sciences* to write abstracts of Russian physiological papers, when I had a bit of spare time," he said. He then did his national service, mostly at the Royal Air Force Institute of Aviation Medicine, where he conducted research on various visual problems and on air-ventilated clothing.

Following his stint in the air force, Brindley returned to Cambridge as a lecturer and then a reader (the equivalents in the United States of assistant and associate professor) until 1968, when he was named a full professor of physiology at the University of London's Institute of Psychiatry. Prior to assuming the position in London, Brindley made a foray to the United States, where he was a visiting associate professor in the Department of Ophthalmology at Johns Hopkins Medical School and in the Department of Physiology and Anatomy at the University of California, Berkeley. Despite his experience in a wide range of medical specialties, Brindley, in his typically self-effacing style, said, "Take my medical career for what it's worth. I'm not a boarded specialist in anything, but I have held a post normally held by people who are board certified in spinal injuries."

Brindley became fascinated with the idea of neural prostheses in 1964 while reading about the work of Otto Foerster and Wilder Penfield, the neurosurgeons who, working separately, experimentally stimulated the brains of awake patients. Brindley was intrigued by the fact that while both neurosurgeons stimulated the visual cortex, their descriptions of their patients' experiences, and the conclusions they drew, differed.

Penfield described the spots of light his patients saw as colored and sometimes complex in shape. Brindley thought Penfield also implied that the spots of light moved in the patients' visual field. Foerster, on the other hand, described the light his patients saw as usually white, small, and fixed.

Brindley felt that Foerster was correct. "His writings gave me confidence, and they fitted very well with observations on defects in the visual field produced by injury to the striate cortex," he said. "These defects are absolutely fixed in the visual field and so I expected if you stimulated the points whose destruction caused a loss in that bit of visual field, then you'll get a sensation in that bit of a visual field, and Foerster's description fitted that idea."

This led Brindley to conclude that "if Foerster is right, and I was confident he was right, then it would be possible by stimulating points on the visual cortex to have an array of electrodes to enable blind people to see." Not one to hesitate, Brindley quickly set about building a system that would stimulate points on the visual cortex. He was able to obtain support for his effort in the amount of about £900, or $1,500 at the time, from the Cambridge University physiology department.

Since electric circuit design was one of Brindley's many hobbies during his early school days and, as he said, "is part of the knowledge of any Cambridge neurophysiologist," he set about designing and making the implant and the external equipment, "mostly with my own hands." Though, he noted, "I did have a fairly unskilled assistant. He was a Cambridge undergraduate, but his subject was Chinese, not science at all. With one unskilled assistant I made this thing."

Describing the process he used to create the visual prosthesis, Brindley made this extremely complex task sound almost simple. "I knew silicone rubber was a promising insulating material, and it seemed common sense to use platinum as the electrode material. Well that was it. There was nothing special about the rest. Naturally, I didn't use seriously toxic materials."

To confirm the biological compatibility, safety, and efficacy of the electrodes he had developed, and to determine safe electrical stimulation levels, Brindley borrowed from previous work and implanted the electrodes he invented into the motor cortices of baboons and rhesus monkeys. Having produced movement by stimulating the motor cortex, Brindley was confident the electrodes would also produce spots of light once implanted in the visual cortex. "We knew how much current was needed to produce a motor response by stimulating the motor cortex in a monkey, for example. And we knew from Foerster and Penfield that the amount of current needed for the visual cortex was nearly the same as for the motor cortex," he explained.

After three years of development and testing, Brindley was convinced that his visual prosthesis was safe and ready for human implantation. To recruit a volunteer, he went to the London Hospital, where he had once been the ophthalmic house surgeon, and discussed the project with his former chief. Brindley recalled him saying, "I think I have a good patient for you. She's a nurse who would think it a thing she would like to do—to be a volunteer for an experimental implant, which would not be useful to her, or at least was very unlikely to be useful to her, but would be the beginning of a development to be useful to other people." The nurse, who was 54 years old, had lost her sight four years earlier to glaucoma and a retinal detachment. Brindley approached her, and as predicted, she agreed to have the experimental implant placed in her brain.

So in 1967, neurosurgeon Walpole Lewin, who had observed Brindley place similar electrodes on monkeys' brains, performed the surgery on the nurse with Brindley at his side at the Addenbrooke's Hospital in Cambridge. The implant consisted of eighty-one electrodes mounted on a single sheet of silicone rubber, which was placed on the visual cortex of the right hemisphere of the brain. Each electrode was connected by a wire to its own diode and induction coil, or crystal radio receiver, placed over the skull but under the scalp so that power and information could be transmitted through the scalp without any connections penetrating the skin—an extremely advanced concept for its time. When a transmitter coil was held up to the woman's scalp, each electrode could be addressed individually. Of the eighty-one implanted electrodes, fifty proved to be operational, and when they were activated, just as Brindley had anticipated, the patient saw fixed spots of light. For each electrode, she saw a different spot in the left half of her visual field. Typically, each hemisphere of the brain provides function to the opposite side of the body.

Dramatic as the results were, they did not prove to be of any practical use to the patient. As Brindley put it, "We never attempted to use it as a getting about device. She was a skilled user of a guide dog, and had no problems in getting about." Nonetheless, since the patient did not work, lived by herself, and had no close relatives, "this gave her an interest in life because we did a great deal of testing, and spent an enormous amount of time on it to map the spots of light accurately," said Brindley.

The following year, Brindley moved from Cambridge to London to take up his position as professor of physiology in the University of London's Institute of Psychiatry. During the same period, the U.K. Medical Research Council, which is similar to the National Institutes of Health in the United States, started a Neurological Prosthesis Unit under Brindley's leadership. Peter Donaldson, an engineer who worked with Brindley at Cambridge, rejoined him at the Neurological Prosthesis Unit, where he went to work designing a second-generation visual cortex implant. The resulting system had one hundred electrodes arranged in rows and columns. The idea behind the newer configuration was to activate various combinations of electrodes to enable the user to see letters. The apparatus was implanted in a blind 50-year-old male patient, and the nurse's original visual prosthesis was replaced with the second-generation system. The results were less than a resounding success. The male patient was able to see spots of light, but they were less clear than the spots seen by the nurse with the original visual implant. Brindley speculated that this might be because the second patient had been blind for many years longer than the nurse before receiving the implant.

In the final analysis, both patients were able to "see" letters, but projecting them into their visual fields was a slow process, and they were able to read Braille using the old-fashioned touch method considerably faster. Eventually, the nurse developed an infection, apparently from the implanted electrodes, which were subsequently removed from her brain, and Brindley's visual prosthesis effort died, but not before it inspired considerable interest among like-minded scientists in the United States.

Spurred on by Brindley's work, Karl Frank, a neurophysiologist at the National Institutes of Health, had initiated an effort to sponsor further development of a prosthetic device to be implanted on the visual cortex. In 1969, he was joined by a young physician named Terry Hambrecht. Shortly thereafter, the two of them formed the NIH Neural Prosthesis Program. The newly formed program set about expanding on Brindley's advancements by investigating a visual prosthetic system of its own based on microelectrodes that penetrate the visual cortex. Members of the NIH group did, in fact, successfully implant a blind volunteer who was able to see spots of light and to recognize simple patterns based on how these spots of light were activated by stimulation. Information developed by the Neural Prosthesis Program on what constituted safe levels of electrical current stimulation are still being used by the U.S. Food and Drug Administration when it reviews implantable neural stimulation devices.

For his part, Brindley said, "it was going to be very difficult and very expensive to get the visual prosthesis useful, if it could be made useful at all, so I made the decision to abandon it. I was happy in abandoning it, because I knew that Hambrecht was carrying on, so the project wasn't going to die unless it deserved to die." And, said Brindley, as he reflected on the visual cortex prosthesis concept over thirty years later, "I now think it probably did deserve to die because I think it's never going to produce things that can't be done better in other ways." Brindley's position is based on the fact that his primary goal in developing the visual prosthesis was to enable blind people to read. Since he invented the implant, character-recognition computer programs have improved greatly, so that machines can now read aloud from the printed page. This development has, in his opinion, usurped his purpose for developing the cortical visual implant. In any event, when he abandoned the project, he thought it had potential only to "be useful to a few rich blind people. Poor people will never be able to have it," said Brindley. So he concluded, "Fine, there are plenty of rich people in America. The Americans are working along these lines, let the Americans do good for their rich blind. There are very few rich blind in England. They can go to America for American implants, there is no need for us to continue." And, in fact, some researchers in the United States are still pursuing a visual cortex prosthesis.

Despite Brindley's misgivings, Hambrecht noted that his investigation of a visual implant had more influence on the field of neural prostheses than any prior event because it had broken numerous barriers. Up to the time the nurse was implanted, there was an unstated taboo on implanting the brain. And though the electrodes were eventually removed, the nurse had them on her brain for a total of some ten years, proving that the body would accept such a device for a long period of time and that the device itself would survive. Brindley had also found that the brightness of the spots of light seen by his implant subjects could be altered by the amount of current passed through the electrodes.

Meanwhile, Brindley turned his attention elsewhere. This time he focused on developing neural prosthetic implants to return movement to people with paralysis.

The first motor prosthetic system Brindley tackled was designed to enable paraplegics to stand and return to a sitting position on their own. This would give them the mobility necessary to transfer from wheelchair to automobile, bed, or toilet and back again without having to rely on assistance from another individual. The system was also intended to enable users to accomplish "swing through walking" using crutches but without having to wear the cumbersome knee braces that were otherwise required. The goal of this line of neural prostheses was "directed towards a much more elaborate kind of implant-driven walking, but at first we thought that at the minimum it would be used to have the patient be able to go from sitting in a wheelchair to rise to his feet and stand up, and sit down again," said Brindley.

This time around, he wanted to make sure his innovation would be successful, so he spent eight years in the laboratory perfecting the concept and conducting animal tests prior to doing the first human implants in 1978. The first recipients of his standing system were two soldiers who were paralyzed by battle-related injuries. When Brindley asked them where they were injured, "both of them said, 'In the back,'" he recalled. "I said, 'I don't mean that. I meant where geographically?' and they said, 'We're not allowed to tell you.'"

The prosthesis they received included six electrodes implanted on the femoral nerves at the groin and the superior and interior gluteal nerves in the buttock. To activate the system, the user had a control box attached to a waist belt, while the transmitters were taped to the skin over the implanted receivers. After the patient flipped a switch to activate the system, there was a delay before power was fed to the embedded electrodes, giving the user time to grasp his crutches. The stimulation came on gradually, so the person could rise or sit slowly, much as an able-bodied individual does.

Both soldiers successfully used the system to stand and sit and to walk without braces, yet according to Brindley, "Neither of them used these things substantially for practical purposes." For them it was easier to slap on their braces than to activate the electronic system embedded in their bodies. Brindley proceeded with a few additional human implants of these devices, but in his words, "They were disappointing." This did not, however, mean that he was not successful, since he had demonstrated that externally generated electrical stimulation could be used to give controlled motion to the paralyzed.

This inspired Brindley's colleague, David Rushton, a professor of rehabilitation medicine at King's College Hospital in London, to take up the motor prosthetic cudgel and develop a lumbar anterior root stimulator, which he implanted in a paraplegic woman. It differed from the system implanted in the soldiers, in that their implants stimulated the peripheral nerves that branch off from the spine to activate the leg muscles. Rushton's system stimulated the nerves right at the spine. With this neural prosthetic, the implantee was able to propel a tricycle using her own muscle power. Other investigators funded by the United Kingdom's Medical Research Council, developed a neural prosthetic implant to allow quadriplegics to move a hand, and, according to Brindley, this device "was used by the patient very much."

Many of Brindley's inventions were precedent setters taken up by other researchers after he first developed them and then moved on to other projects. In other cases, simultaneous efforts were going on elsewhere in the world to develop similar neural prosthetic devices. Such was the case with a hand manipulation system being developed by a group of biomedical engineers at Case Western Reserve University in Cleveland, Ohio, who later coalesced into the Cleveland FES (Functional Electrical Stimulation) Center. As with the visual prosthesis, Brindley was comfortable with the fact that a hand control device was being pursued elsewhere and decided to move on to the next big thing in neural prostheses. Though he did note that "our technology was simpler, and from that point of view might have been cheaper. It certainly was cheaper to develop, and it might have been cheaper to produce. That would have been, from the British point of view, very important, however it didn't go any further."

Although the Cleveland FES Center and the Medical Research Council were undertaking parallel efforts, their work was totally independent of each other. Their comparable advancements were due to the fact that advancements in microelectronics and materials science made the time ripe for the creation of neural prostheses. As a result, "it was not surprising it got done in the two centers in the world that were equipped to develop it," said Brindley.

Meanwhile, Brindley continued in what had become his modus operandi —creating a whole new field of implant technology and leaving its perfection to others while he moved on to his next invention. This time, he turned his attention back to the brain. But rather than sending signals in, as he had done with the visual prosthesis, he focused on recording signals from the motor cortex, where the electrical signals, or action potentials, that activate muscles are created. His goal was to capture those signals and use them to drive motor prostheses, like the hand manipulation device. If this were successful, a completely paralyzed individual could theoretically think an implant into motion, much as an able-bodied individual activates a healthy limb or digit.

Brindley's work involved implanting electrodes in the motor cortex of a baboon and recording signals as it moved a hand to pick up a reward. He was able to isolate the signals related to that movement from the thousands upon thousands of other neural firings going on at the same time. As with many of his other endeavors, this was some of the earliest work in the field, yet at the time he decided to abandon this line of investigation for several reasons. One was a concern over inducing epileptic seizures in implantees. "If you mess about in the motor cortex you're liable to get a preepileptic focus," he said. "You can do so by messing about with any of the cerebral cortex, but the motor cortex is particularly bad from this point of view, so we thought it was rather risky, and as there was not a great deal of gain from it, we didn't pursue it."

Economics was another reason for abandoning this line of research.

Funding for the United Kingdom's Neurological Prosthesis Unit was limited, and Brindley did not want to undertake the pursuit of projects that he felt were questionable in their ability to help the majority of disabled people. So he turned his attention to other devices he thought would be less expensive and more practical in the short term. Thus the bladder, bowel, and sexual stimulator was born. A basic three-channel device, it was developed primarily to enable paralyzed patients to void urine regularly and completely without the need for catheterization. The goal was to eliminate the frequent urinary tract infections and more serious renal failure that was, and to some extent still is, all too common among people who are paralyzed. As a bonus side effect, Brindley found that he could use the same device to induce bowel movements, another problem for individuals with paralysis, and in some men activating the stimulator also produced erections. The device was commercialized in 1982 by FineTech (now FineTech Medical Ltd.), a British general engineering company. The system is sold throughout Europe, Australia, New Zealand, and Singapore as the FineTech Brindley Bladder Control System. In the United States, it was approved for widespread clinical use by the FDA in 1998 and is sold under the Vocare name. It has since been implanted in thousands of patients (see chapter 8).

Brindley also commercialized an electrically activated neural prosthetic device to enable men with spinal cord injuries to emit semen. Known as the hypogastric plexus stimulator, this device activated the sympathetic nerve fibers supplying the seminal vesicles and prostate gland. It was one of several of Brindley's inventions that he commercialized through a company he set up with his wife in 1990. During its ten-year tenure as a commercial product, the semen stimulator was implanted in a substantial number of men, many of whom were able to father children thanks to Brindley's implant. The semen system expired when Brindley closed his company down because of Britain's National Health Service requirement that medical equipment manufacturers be insured for up to £10 million per claim, "and we couldn't afford the insurance," he said. In retrospect, even "if somebody said, 'What about reintroducing the hypogastric plexus stimulator?' I'd say, 'it's not worth it.' There are enough other ways to get sperm to fertilize eggs without this." Those other methods include using a vibrator to produce reflex ejaculation, or activating an electrode temporarily placed in the rectum in a method called, appropriately enough, electroejaculation.

As well as being a leader in the development of neural prostheses, Brindley was well ahead of his time when it came to the use of drugs to deal with erectile dysfunction. In addition to the injection method he graphically demonstrated in Las Vegas, during the mid-1980s he designed an apparatus that is implanted in the scrotum between the testicles. Rather than being electrically operated, this system consists of a mechanical pump and reservoir that contains the drug sodium nitroprusside, which is used by anesthetists to lower blood pressure and which Brindley selected because it is inexpensive and works well in inducing erections. Brindley had one of these devices implanted in himself in 1992, and twelve years later was still using it. He also produced and commercially sold the system through his company.

When the user wants to achieve an erection, he simply uses his fingers to actuate the pump embedded in his scrotum that releases the muscle relaxant into the erectile tissue in the penis known as the corpora cavernosa. This causes the arteries that feed into the erectile tissue to dilate, increasing the blood flow to the area that creates an erection. Depending on the frequency of one's sexual activity, a reservoir full of the chemical is good for about three months, after which it must be refilled by injecting additional chemical through the skin of the scrotum into the reservoir.

A major challenge when designing the erection system was finding a material that could withstand being repeatedly penetrated by the injection needle without leaking. Unable to locate any, Brindley simply concocted his own inert silicone membrane to be used as the implanted reservoir material. While developing the membrane, Brindley tested it by puncturing it with a syringe. "After three hundred times, I got bored and didn't puncture any more," he said. He has personally refilled his reservoir some fifty times, and the system continues to work well with no manifestation of leakage.

———

BRINDLEY ALSO TURNED HIS CREATIVE SPIRIT to music, another of his loves. In doing so, he invented a new musical instrument he calls the Undilector, of which there is only one in the world. Brindley plays the Undilector in amateur concerts and writes music specifically for it. And, he said, "I'm trying to persuade better composers than myself to write for it."

As with all of his inventions, the Undilector came into being as the result of a physical need. Brindley used to play the violin, viola, and bassoon, all of which he had to give up as age affected his ability to manipulate his fingers. That left him with the French horn, which he has played for some time and can still handle. But he wanted more, so he invented it. The Undilector is a computer-controlled wind instrument that Brindley designed to resemble a bassoon in size, weight, and feel, though he placed the switches, which replace conventional valves, where he can conveniently reach them. He derived the instrument's name from the Latin words *unda*, which means wave, and *lector*, a person who reads aloud, since the instrument reads sound waves aloud.

The Undilector stands on the ground. The performer sits to play it and blows into it, much like a bassoon. The fingering is also similar to that of a woodwind, but the Undilector also has switches, which are operated by the knuckles and left knee. They are used to turn notes into cords, create vibrato, or change the timbre of the music. As the Undilectorist blows, a pressure sensor is activated. As with any wind instrument, the harder one blows into it, the louder the sound, except that in this case a computer is involved. The waveforms of the sounds are programmed into a microprocessor that produces them at varying speeds, depending upon how the player manipulates the keys of the instrument. If the waveforms are read quickly, a high note results. Conversely, slow reading produces a low note. And though the Undilector sounds like a bassoon, it can also sound like other instruments if Brindley wants it to, simply by reprogramming the waveforms in the computer's memory.

As for how this instrument relates to other computer-generated music, Brindley takes the same approach as he has with his neural prosthetic innovations. He created it independently of what others were doing and was not particularly interested in their efforts. When "the time is ripe, lots of different people have ideas and they converge, but I'm not really interested in that," he said. "I made this instrument for the pleasure of playing it. And, it's a good instrument in my judgment."

When not inventing neural prostheses or musical instruments, Brindley has also been a long-distance runner, having run seven marathons. At age 62, he won a bronze medal as a member of a 4 × 400 meter relay team at the World Veterans Athletic Championships held in Melbourne, Australia. But now, he said, "I only run to catch buses."

His mind, however, continues to run at full throttle.

5

Accidental Pioneers

In one of those twists of fate that no one can foresee, a tragic 1959 automobile accident that left a young college freshman unable to move his arms or legs was to later provide a major impetus to the development of neural prostheses and thereby give hope to thousands of other quadriplegics just like him.

The victim of the accident was Jim Rider, a student at the University of Iowa at the time. Rider hailed from the small farming town of Galesburg, Illinois, where he and Terry Hambrecht had been high school buddies. Hambrecht was a sophomore studying electrical engineering at Purdue University in West Lafayette, Indiana, when the accident occurred. Deeply troubled by his friend's fate, Hambrecht agonized over how he might apply the knowledge he was obtaining as a budding electrical engineer to help his friend. He knew that Jim's muscles were intact and could still function if only the electrical signals needed for them to contract could get to them. But that was impossible since the nerves that traveled through his spine and carried signals to the muscles were irreparably damaged. Yet Hambrecht wondered if he might apply electricity to his friend's skin right at the site of his paralyzed muscles to get them to move. This was essentially the same thing Benjamin Franklin attempted approximately two hundred years earlier.

Hambrecht asked Rider if he would mind if he conducted a few experiments on his muscles. "No, they're no good to me," Hambrecht recalled his friend replying. So he set about designing and building an electrical muscle stimulator that would allow him to apply electrical current to Rider's muscles. "I put some surface electrodes on the outside of his muscles and ran some electric current through them, and, of course, they contracted but in a terribly uncoordinated manner." This was not unlike the results Franklin experienced when he attempted his experiment. But Franklin did not have access to the electrical technology that was at Hambrecht's disposal, so he took his experiments no further.

Hambrecht, however, was able to contemplate doing much more with his idea. "There's really no reason why we can't put together a control system and have the whole thing implanted and use it to make the muscles contract in a coordinated manner," he thought, and resolved to attempt to do just that. But to do so would require that he learn a lot more about electricity and the human body. "Because I was in engineering school, I realized what I had now was crude muscle contractions; what I needed was very careful control to be able to control those contractions in a functional way. It was at that point I decided I would try to concentrate biology and medicine with engineering to see if I could learn enough to do it in a very constructive and functional way," said Hambrecht.

So after receiving his bachelor's degree from Purdue, Hambrecht went on to the Massachusetts Institute of Technology, where he obtained a master's degree in electrical engineering in 1963. But that still wasn't enough. While working toward his master's, Hambrecht realized that in order to develop a viable system, he needed a medical education as well. So following graduation from MIT, he went on to Johns Hopkins University Medical School, where he received his M.D. in 1968. He then spent a year as an intern at Duke University Hospital. The drive still strong to help Rider and others like him, Hambrecht had a special interest in surgery. "I wanted to see what the real problems were rather than what I thought the problems were, of people with various disorders, and I also wanted to see what techniques were available for implantation," he said.

Meanwhile, the Vietnam War was raging, and it loomed large in the future of all young medical school graduates. "They told us when we were seniors in medical school, even if we were in wheelchairs or on stretchers, we would go to Vietnam," said Hambrecht. Fortunately, he was able to obtain a commission in the Public Health Service, which was considered one of the uniformed services, and he was assigned to the laboratory of Karl Frank at the National Institutes of Health. Frank, who was world renowned for his basic physiological research into the spinal cord, had been inspired by the work of British physician and neural prosthetic pioneer Giles Brindley to develop a prosthesis to restore sight to the blind by implanting electrodes on the visual cortex of the brain. That was in 1969, and Frank's work fell under the aegis of the NIH Sensory Prosthesis Program. Starting out as the equivalent to a second lieutenant in the U.S. Public Health Service, Hambrecht intended to stay at the NIH for only two years. Thirty years later, he retired from there with the rank of colonel.

Unfortunately, by the time Hambrecht got to the NIH the friend he was so intent on helping had died of a gastric ulcer he didn't know he had because of his inability to feel pain. "There are so many complications one can get as a quadriplegic, and at the time care wasn't nearly as good as it is now," said Hambrecht. "The only experiments he participated in were the ones I carried out on him, which were pretty crude at that time. It's too bad, because I would have loved him to get one of Hunter Peckham's systems in him," said Hambrecht, referring to a device, developed by another neural prosthetic leader, that enables quadriplegics to manipulate a hand (see chapter 6).

Despite his friend's death, Hambrecht was intent on developing prosthetic devices to help other paralyzed individuals benefit from the implantation of manmade devices to interact with the nervous system. Toward that end, he lobbied hard for the NIH to expand the Sensory Prosthesis Program to include paralysis in addition to sensory disabilities in its research endeavors. His campaign was given added impetus by the fact that the technology being developed at the NIH for the visual cortex prosthesis could be applicable to other types of neural prostheses as well, all of which involve direct connection to the nervous system. It took two years for Hambrecht's efforts to come to fruition, but eventually he was successful in getting the NIH to launch an all-inclusive program. In keeping with the expanded emphasis, the Sensory Prosthesis Program was renamed the Neural Prosthesis Program, and Hambrecht was named the co-director. Three years later, Hambrecht took over the sole directorship. In 1986, he was joined by William Heetderks as his deputy. Heetderks took over leadership of the Neural Prosthesis Program when Hambrecht retired in 1999.

To spur the development of neural prostheses, one of Hambrecht's early tasks at the NIH was to pose a series of questions that he felt had to be answered before neural prostheses could become part of the medical arsenal to overcome sensory and motor deprivation. "The biggest limitations were in the biomaterials field," said Hambrecht. "Very little was available, and very little was known about the body's long-term reactions to the materials that were available, and whether the materials themselves would hold up in the body." A second major line of questioning pertained to the development of miniature electrodes. Prior to the formation of the Neural Prosthesis Program, electrodes were little more than very fine wires that could be placed in the body. Precious little had been done to create the multielectrode arrays that would be required to provide sophisticated control or to record from multiple neurons. Yet another area in which too little was known was how tissue reacts to artificial stimulation, how much stimulation is required, and how much is too much?

Hambrecht's program raised the questions, obtained NIH funding, and encouraged contract proposals from those scientists who were best able to answer those questions. "Our program was the overseer," said Hambrecht. "We never claimed to have designed the experiments. We raised the questions that we knew had to be answered and then we put out advertisements for people who might want to bid on these contracts with ideas of how they would go about experimentally answering these questions. That was the basis of the Neural Prosthesis Program." In addition to this extramural part of the program, the NIH continued with some of its own in-house research, such as the ongoing visual cortex prosthesis investigation, which included developing microelectrodes and working with patients.

Another major thrust of Hambrecht's efforts was to get researchers from various fields who were working on the development of neural prostheses to communicate with each other and to do so in the same language. Virtually every field of scientific endeavor has its own lingo, and for the unanointed, understanding what those within the fold are saying can be a daunting task. Hambrecht aimed to make it less so, since if neural prostheses were ever to become a widespread reality, it would require the cooperation of physicians, engineers, chemists, materials scientists, computer experts, and many more. "Everybody had a different language. The electrode physiologists couldn't talk to the electrodochemists, and they couldn't talk to the anatomists," said Hambrecht.

To open lines of communication among these neural prosthesis researchers, the newly formed Neural Prosthesis Program convened the first annual neural prosthesis workshop in 1970. "What we did was convince everybody to define their buzz words right from the beginning, when they started to give a talk, and to try not to talk as though they were talking to colleagues in their specific fields," said Hambrecht. To further mutual understanding, contract recipients were required to produce quarterly reports about the progress of their work in what Hambrecht termed "*Scientific American* style English, so others could understand it. When we got together for workshops we assumed people had read the progress reports and were up to date so the speakers could talk about their most recent research at a level that was truly progressive rather than just didactic teaching." That workshop has been held every year since and has grown exponentially in the number of attendees.

ONE OF THE RESEARCHERS WHO CAME KNOCKING on the door of the Neural Prosthesis Program looking for funding was J. Thomas Mortimer, who headed a young program at Case Western Reserve University aimed at using electrical stimulation to return movement to the paralyzed in a field now known as functional electrical stimulation (FES). This is precisely what Hambrecht had wanted for his now deceased friend.

Like Hambrecht, Tom Mortimer, who was born in 1939 and grew up in Amarillo, Texas, had a close high school friend who was paralyzed in an automobile accident. Mortimer's friend Joe Macrander became wheelchair bound during the summer between their tenth and eleventh grades. And as with Hambrecht, the misfortune of Mortimer's friend became a significant factor in the direction of his career. "It was something I paid attention to," said Mortimer.

Mortimer's path to neural prosthetic research was, however, considerably more convoluted than Hambrecht's. As a teenage product of 1950s northern Texas, Mortimer was almost obligated to do a stint as an auto mechanic. "Cars were everything," said the lanky, laid-back Mortimer, sporting a Texas-sized belt buckle, as he spoke in his Case Western Reserve office some fifty years later. "I figured out I could leave school at noon, and go work on cars, and make money," he said of the trade high school program he opted to follow. After high school graduation and two years as a mechanic in a Chrysler dealership, Mortimer decided he wanted more. His dream was to become an engineer and design things rather than repair them. Though his high school academic record was less than stellar, he applied to Texas Technological College and was accepted. His auto mechanic's skills stood him in good stead during his college days, since in his words, "even in Texas, with tuition fifty dollars per semester, our family didn't have that kind of money. But as a mechanic I could earn enough in the summer to put myself through a full school year."

When it came time to select a specific engineering major, Mortimer was at a loss, so he wandered the Lubbock campus of Texas Tech looking at engineering buildings. "I remember walking by the mechanical engineering building, and that sounded like automobiles. I didn't think I wanted to do that. Textile engineering was the next building. They didn't seem to have enough status in the world, and civil engineering was kind of old. Then I went by chemical engineering, and that sounded interesting because I had a Gilbert chemistry set as a kid. Then there was the electrical engineering building, and I thought, 'These guys seem to have status.'" So he decided to become an electrical engineer, because they are "the elite engineers . . . That was the rationale," said Mortimer in the folksy Texas twang he retains.

The transition from auto mechanic in Amarillo to engineering student was not an easy one for Mortimer, who admits to having "had a rather speckled career as a student. My last high school math class was plane geometry. I got an F in that because I wasn't paying much attention, and my last science course was in the eighth grade." So in something of an understatement, Mortimer, now a semiretired professor of biomedical engineering and a noted name in the annals of neural prostheses, said, "College was an eye-opener for me. I had a lot of remedial work to do. It wasn't clear that I was going to be able to finish the first year." But the thought of the embarrassment of flunking out of college and going back to the auto dealership drove him to greater heights.

Mortimer traveled a long intellectual road during his undergraduate days, to the point that during his senior year at Texas Tech, he decided he wanted to go on to graduate school. He had heard about a new up and coming field called biomedical engineering that tweaked his interest because of what had happened to his high school friend. At the time, there were not a lot of universities that offered programs in biomedical engineering, but the Case Institute of Technology (now Case Western Reserve) in Cleveland was one that did. Coincidentally, the dean of Texas Tech's college of engineering at the time was a graduate of Case, so thinking a letter of recommendation from him might help, Mortimer applied there.

Because of his stated interest in helping paralyzed people, Mortimer's application found its way to the desk of Jim Reswick, the head of what was then called the Engineering Design Center. Reswick was leading a team of students in the design of an exoskeletal apparatus intended to enable quadriplegic patients to move their arms. The device was designed to be strapped onto a paralyzed patient rather than implanted in the patient's muscles. Reswick offered Mortimer an assistantship with a stipend of $135 per month. Mortimer was delighted with the offer. So in January 1964, following his graduation from Texas Tech, he piled his few possessions into his 1948 Ford and headed north to Cleveland to begin his graduate student career. Little did he know that he would still be there over forty years later.

Within a few weeks of Mortimer's arrival at Case, another man attracted by the work being done there also arrived on the scene. His name was Lojze Vodovnik, and his trip was a bit longer than Mortimer's. He had come from Slovenia, where he had recently received a doctorate in electrical engineering from the University of Ljubljana. Vodovnik became interested in the electrical stimulation of paralyzed muscles through articles he had read about the work of researchers at Cleveland, and he had been accepted as a postdoctoral fellow at Case. As luck would have it, he and Mortimer landed in the same housing complex and both went to work on Reswick's exoskeletal device. Since Vodovnik did not have a car, Mortimer gave him rides to and from the laboratory, and in the process the two electrical engineers became fast friends.

Reswick's project, known as the Arm-Aid, was an awkward computer-controlled, pneumatically actuated device that when strapped on gave the wearer a robotlike look. Eventually, it became apparent that the device was too ungainly to be practical. Meanwhile, Canadian researcher John Basmajian had developed an electrode that could be placed inside the body to record electrical activity within muscles. Inspired by the recent success of the cardiac pacemaker, researchers at Case decided to adapt Basmajian's implantable electrode concept to develop electrodes that would pump electricity into muscles rather than record from them. "What was in the minds of most of us," said Mortimer, "was if you can do it in the heart, you can do it in the muscles." Though Case was not the only place where scientists were working to achieve coordinated movement of paralyzed muscles, it emerged as an international leader in the field of functional electrical stimulation. "What distinguished us from all the other people that were doing this was our willingness to go inside the body. When everybody else stayed on the surface, we went inside," said Mortimer. And, he added, "we began to pay a lot of attention to the tissues, the muscle, the physiology, the life science aspect of it."

When Mortimer spoke of going inside the body, it was not just the bodies of paralyzed people he was speaking of. Before electrodes could be permanently implanted in patients, their safety and efficacy in humans had to be proven, and what better way to do so than for the researchers to test their inventions on themselves? If they were confident enough to do so, then the product of their research must be good. As a result of this philosophy, many an older neural prosthetic researcher still carries shrapnel from battles at the laboratory bench. Such self-experimentation became less practical following the passage of the Medical Devices Amendment in 1976, which requires FDA approval prior to the human implantation of neural prostheses.

One of the early researchers to lay his body on the line was Vodovnik, who in 1965, just one year after arriving at Case, had a radio transmitter

implanted in his right shoulder. A receiver and electrical stimulating electrodes were implanted in a dog's hind leg. The transmitter was activated by
the movement of Vodovnik's shoulder. When he voluntarily raised his shoulder, the electrodes in the dog's leg were activated, and it involuntarily raised
its hind leg. Vodovnik thereby demonstrated that a muscle under voluntary
control could be used to drive another muscle that was not. This proved that
if a transmitter were implanted in a person with paralysis where he or she
retained voluntary movement, that movement could be used to actuate electrodes implanted elsewhere in the body.

During this period of wide-ranging experimentation, Mortimer turned
his attention to learning how to record from muscle spindles, or stretch
receptors. These receptors activate sensory nerves that send information
about the length of the muscle to the spinal cord. This information is used by
motor neurons that send electrical impulses back to the muscle to coordinate smooth, efficient, spontaneous contractions sufficient to accomplish
the task at hand. This closed-loop system is how you can feel the difference
between lifting a glass of beer and a barbell. And because the processing is
done in the spine rather than the brain, the signals are processed quickly
enough to enable you to grab a slipping glass or instantaneously pull away
from a hot surface.

Individuals with paralysis whose muscles are activated by implanted
electrodes do not have this natural feedback capability. Because they cannot
feel the pressure of what they may be holding or lifting, the only sensory
feedback they have is visual. The thrust of Mortimer's work was to enable the
development of an implantable closed-loop system that would record from
the functioning sensory receptors in the muscles and send the information
to the spinal cord above the damaged area. In essence, it would act as a
bridge from the muscles to the spine allowing the natural signals from the
muscles to leap over the damaged nerves. Thus, a hand grasp system, for
example, would automatically tighten the user's hand around a slipping
glass or hold a Styrofoam cup without crushing it, just as an able-bodied
individual instinctively does.

The first step in accomplishing this lofty goal was for Mortimer to learn
how to record from the spindle fibers in the muscles and to build an in-

strument that could extract the recorded information. To do so, he began experimenting on cats, a task that as a trained electrical engineer with no experience in such work, he found distasteful.

"I got very familiar with the cat lower leg. How to open up the spinal column, how to record from the spinal roots, how to measure muscle forces, and record neural information," said Mortimer. Cats were the animal of choice for his experiments because they were most often used for neuromuscular experiments and there was some technical literature with which he could compare his results. Yet Mortimer found these experiments "just awful." His objection was not ethical but practical. As an electrical engineer, he found transistors to be more predictable than cats. He could turn on a transistor, go to lunch, come back, and the transistor would still be the same. That was not so with cats. "If you start an experiment on one in the morning, you're lucky if you finish by the next morning," he said.

He so disliked this work that for a period of time he considered abandoning biomedical engineering in favor of oceanography. But the goal of helping the paralyzed, as well as some coaxing from his advisor Reswick, drew him back into the fold. Reswick asked Mortimer to go to work with Norman Shealy, a neurosurgeon who was working on developing a dorsal column stimulator, a device to suppress pain using electrodes attached to the back of the spinal cord. Reswick assured Mortimer that although this work would entail more animal experiments, he did not have to worry; Shealy would take care of that end of it. "Well," said Mortimer, "it wasn't true. I ended up working in the brain trying to understand how much stimulus we could put in before we injured the tissue."

The upshot of this work was a doctoral dissertation for Mortimer and the creation of an epidural pain stimulator. This pain mitigation system, which is still being implanted in patients with otherwise uncontrollable pain, operates on the principle of producing something akin to an electronic mosquito bite. By activating specific nerves, it creates mild discomfort, thereby masking the more significant pain. When the Neural Prosthesis Program was formed a few years later, it did not consider pain stimulators to fit within its purview. Because, according to its co-director Hambrecht, "we could not figure out ways to quantify it, and we wanted to be objective about anything we did. It wasn't because there wasn't a big need—there still is." Mortimer felt the same way. After finishing his doctoral dissertation, he said, "I don't want to work on pain because I can't measure it." Nonetheless, according to Hambrecht, the pain stimulator "was a major contribution."

Still trying to get away from animal experiments, Mortimer decided to accept a postdoctoral position at the Chalmers Institute of Technology in Gothenberg, Sweden, where he looked forward to doing "pure engineering kind of stuff. No more pain, no more animals. I was just going to take the signals from a muscle and process them."

The Swedish researchers whose laboratory he joined were studying muscle fatigue as it related to ergonomics. They were working under a contract with local shipyards, where management wanted to understand what caused workers to tire. During their early research, they found that when muscles fatigue, the energy in the electrical signals within the muscles drops in frequency. They were studying this phenomenon to determine how it related to muscles tiring and if that information could be used to create better working environments. Mortimer was brought in to apply his engineering skills to analyzing signals others obtained from animals' muscles. But once he got to Sweden, that changed. In short order, he found that he was not enthralled with the computer programming and the analytical work he had coveted. In the final analysis, unbeknownst to him, he had become somewhat addicted to the lure of hands-on research. With this discovery, Mortimer had something of an epiphany when he realized that as an engineer who was also adept at conducting animal experiments, he had a rare combination of skills that he was now eager to use to further the field of neural prostheses.

At the same time, he became interested in work being done at several other laboratories around the world that resulted in the discovery of two different types of skeletal muscle fibers—skeletal muscles are those that execute voluntary movements, as opposed to heart muscle, for example, which contracts automatically. It was found that there are two basic types of such fibers. Slow-twitch muscle fibers—also called red muscles because of the high levels of oxygen they retain—are built for stamina as opposed to strength. Fast-twitch fibers, on the other hand—or white muscle, which makes up the white meat of chicken—provide more strength but tire quickly and take longer to restore their energy supplies. Those good at endurance sports, such as long-distance runners, have a preponderance of slow-twitch muscle fibers, while sprinters and weight lifters have more fast-twitch muscle fibers.

The more Mortimer read about slow- and fast-twitch muscles, the more he became convinced that this discovery held the key to one of the most vexing problems facing those working on electrical stimulation of paralyzed muscles, namely, that muscles stimulated by means of electrodes fatigue more rapidly than do naturally activated muscles.

Muscle fibers are grouped into motor units, a motor unit being a single nerve fiber and all the muscle fibers it innervates. When a person moves, smaller motor units, which are made up of slow-twitch fibers, are the first to be activated. The larger, fast-twitch motor units that tire more rapidly, come online after the slow-twitch muscles. Mortimer thought that the reverse might be true when muscles are artificially stimulated by electrodes. His theory was based on the fact that the larger fast-twitch motor units have a lower threshold for electrical activation, thus they may be the first to respond when a jolt of externally generated electricity hits. Since slow-twitch fibers are more resistant to electrical stimulation, they might react after initial stimulation. As a result, Mortimer thought, FES devices might circumvent the body's natural ability to stave off muscular fatigue, since the infinitely more sophisticated natural body mechanisms more finely control the muscle activation.

The task of testing Mortimer's hypothesis and finding a way around this reverse activation conundrum, if it was correct, fell to a young graduate student named P. Hunter Peckham, who was the first Ph.D. candidate Mortimer advised. Peckham, originally from Elmira, New York, had come to Case Western Reserve from Clarkson University in Potsdam, New York, in 1966, immediately after receiving a bachelor's degree in mechanical engineering. His arrival at Case resulted from his having casually picked up a technical magazine to browse through while waiting to see his undergraduate advisor during his senior year at Clarkson. Peckham, who was soon to become a rising star in the young field of neural prostheses, did not know what he wanted to do after graduation and was at his advisor's office to discuss options. While perusing the magazine, he came across an article about implantable heart valves that grabbed his attention. "It never occurred to me that you actually could make things to go inside the human body," said Peckham from his vantage point as founder and director of the Cleveland FES Center some forty years later. This, coupled with the fact that heart valves involve fluid flow, something mechanical engineers are intimately familiar with, inspired Peckham to think about going to graduate school to study biomedical engineering. He found that Case was one of the few schools to have such a program, so he applied and was accepted.

When Peckham arrived at Case, he planned to study fluid flows through the body, but that was a short-lived goal. Not long after his arrival, he learned that Mortimer and company were electrically stimulating muscles. " 'Bingo,' I thought, 'that's for me,' " said Peckham, who still maintains that studentlike enthusiasm for his chosen field. He eventually became a protégé of Mortimer's, and was given the task of studying muscle fatigue, a subject that became his dissertation topic.

As Peckham began to tackle the subject, many experts predicted that functional electrical stimulation would never work, precisely because paralyzed muscles fatigued very quickly and, they said, could never generate enough force to be functional. At the time, "both of those things were true," said Peckham. Building on Mortimer's hunch about what caused artificially stimulated muscles to fatigue so quickly, Peckham found that whatever the cause was, it was more than what one would expect of muscles that had not been used for some time. Atrophy makes muscles weak, but in the case of paralyzed muscle, he discovered that there was also a change in metabolism that made them unable to sustain contractions for long periods of time. He then was able to establish that the problem was indeed caused by reverse activation, just as Mortimer had suspected.

Peckham's next step was to set about finding a way to overcome the reverse stimulation problem. Ironically, he found that this could be done using the cause of the problem, namely, artificial electrical stimulation, as the cure as well. Just as athletes improve strength and endurance by exercise, so too, Peckham discovered, could electrical stimulation be used to work paralyzed muscles in a precise manner that would induce them to reverse course and become less prone to fatigue. The prescription involved a complex combination of stimulation rates and sequences. With the major hurdle of muscle fatigue conquered, functional electrical stimulation came considerably closer to reality.

Today, anyone who has ever had physical therapy following surgery or muscular injury and has experienced the tingling sensation created by external electrodes applied to their skin to exercise the muscles below has benefited from Peckham's insights.

Peckham received his doctorate based on this work in 1972 and hoped to continue his endeavors at Case as a postdoctoral fellow, but Mortimer had run into a funding crisis. In fact, money was so tight that Mortimer feared for his own job, thinking that if he did not receive funding soon, his program would be closed down and he would be out the door. It was that concern that led him to visit Hambrecht and Frank at their newly formed Neural Prosthesis Program offices at the NIH in Bethesda, Maryland.

Considering what was at stake, Mortimer was more than a bit nervous about the visit but was immediately put at ease by Frank, whom he found to be a friendly, white-haired father figure. Mortimer felt honored by the fact that a man of Frank's stature was familiar with a paper he had written about brain tissue reaction to electrical stimulation. When Frank asked about his funding requirements, Mortimer told him that one of his most pressing needs was to keep Peckham on and to do so he needed $25,000. Frank said he felt that was a doable figure and asked Mortimer to submit a proposal, which he did. The money was granted, and both Mortimer and Peckham stayed on. Since those early days, the NIH program has provided millions of dollars to the neural prosthesis program at Case.

During the early phase of his work, Peckham used stimulators that were placed on the surface of the skin, but it soon became apparent that stimulation would be more effective if the electrodes were implanted so they could move with the muscles. Additionally, external electrodes were cumbersome, and if functional electrical stimulation was ever to become a reality, users would want something more cosmetic. So Peckham and his colleagues turned their attention to creating implantable electrodes. "That was pretty outrageous thinking at the time, that you could actually make a stimulator that went inside the body," said Peckham. Nonetheless, they designed electrodes that passed through the skin directly into muscle tissue, and in 1973, were able to use them to successfully induce controlled movement in the hand of a quadriplegic. But the electrodes were primitive and lasted for a maximum of only four months before needing replacements. Yet the viability of using implanted electrodes to give controlled movement to the paralyzed had been established.

In 1975, the group headed by Mortimer that was responsible for this work took on the name Applied Neural Control Laboratory. Two years later, Peckham formed the Rehabilitation Engineering Center, which gravitated toward a clinical approach, working more with patients. Both organizations were under the aegis of Case Western Reserve University. In 1990, the two entities merged to form the Cleveland FES Center, with Hunter Peckham at the helm. The center, a consortium of Case Western Reserve, the Louis Stokes Cleveland VA Medical Center, and the MetroHealth Medical Center, has become a worldwide focus of neural prosthetic activity aimed at returning movement to people who are paralyzed.

In addition to education, a driving force behind the formation of the Cleveland FES Center was the desire to bring the esoteric FES systems out of the laboratory and into the hospital to be implanted widely in patients who could benefit from them. Peckham recalled an orthopedist saying to him shortly before the formation of the FES Center, "You guys have a great reputation for the science, but why is it that you haven't been able to translate that into helping my patients?" That made Peckham think long and hard about what it would take to do just that. He concluded that he would have to create a multi-institutional atmosphere that went beyond the hallowed halls of academia to work directly with patients. Such an organization, he felt, would also encourage scientists, who are sometimes prone to fine-tune their inventions for too long, to turn them over to others who know more about the real-world marketplace.

The steps involved in doing so are numerous and complex. They involve understanding how and when nerves fire when they do, developing hardware such as connectors, pulse generators, leads, and electrodes to interface with the nerves, and then creating the technology to bring the various pieces together into a system that is programmable by a therapist working with the patient. Then the system must be brought into the clinic. As Peckham put it, "When we design a system it has to be conceived in a way the makes the technology deployable, deliverable, serviceable, and frankly it has to be reimbursable." Thus, reducing the science to practice is, in Peckham's understated words, "very hard."

Students are frequently the font of new ideas for components or entire systems that may be destined to become commercial products. They develop and test their concepts, first using computer modeling, then animal testing to prove their feasibility. In the process, the students cannot help but be infected by Peckham's continuing sense of excitement for the work of neural prostheses.

"It is as captivating today as it was when I arrived at Cleveland in 1966," he said some forty years later, as he listened to a graduate student present his research on a new FES concept. Peckham reveled in the intellectual give and take as faculty members and other students fired questions at the presenter. "You can see all the thinking that's going on. For me it's so exciting, it's going through my veins. I just get chills thinking about it," he said. Peckham also takes delight in how far the field has come since he was a graduate mechanical engineer who was going to study fluid flow through closed vessels. Today's students, said Peckham, "are a lot smarter." They come to Cleveland knowing about FES. "We actively work to identify and recruit students who want to work in this neural field, and we get the very best students. They are stupendous."

Once students, faculty, and staff develop new FES systems, they must be tested in human volunteers who are willing to put their bodies on the line so that the researchers can perfect the systems and turn them over to commercial companies. One of the first patients to do so was James W. Jatich, a quadriplegic who helped develop the Freehand, an implantable FES system that enables quadriplegics to manipulate a hand so they can feed and groom themselves and even operate a computer. It became the first commercial product to flow from the Cleveland FES Center.

6
Giving a Hand

H is real name is Jim Jatich, but he might well be called Mr. Electrode, since he has probably had more electrodes implanted in his body than any human in history—by his count, 150 to 200, with a maximum of 20 at any one time. This has involved approximately ten surgeries over more than twenty-six years, and still counting. And Jatich is almost jovial about it. Rather than an imposition, an inconvenience, or a painful experience, he considers it something of a privilege to serve as a virtual pin cushion for a team that designs, builds, and implants a variety of electrically actuated devices to return movement to people like himself.

His most recent count is sixteen electrodes divided between two implanted functional electrical stimulation systems—one in each hand—which until recently enabled him to manipulate both of his paralyzed hands. One of the systems failed after being in place for five years. Jatich could have had it replaced immediately but decided to wait for the next generation. But even with one operational hand, he can feed and groom himself and operate a computer, which has allowed him to run a home-based drafting business utilizing the skills he developed as a junior engineer before his accident.

The fateful day that led Jatich to become a champion of functional electrical stimulation was August 28, 1977. Jatich, who was 28 years old at the time, and a few friends had been painting a house in Akron, Ohio, that he was getting ready to sell. They had worked late into a hot, muggy summer night and decided to cool off with a swim in a nearby lake after calling it quits for the day. "About five of us dove into the lake," said Jatich. "I was the last one to dive in, and I hit something. It didn't knock me out; I just was stunned and sank to the bottom, my face in seaweed. I tried holding my breath as long as I could. Luckily some fishermen saw where I went in and didn't come up, and they told my friends to come help me."

After bringing him to the surface, his friends wrapped a towel around his neck and held him as still as they could until the volunteer fire department showed up and took him to a nearby hospital. The diagnosis was not good; Jatich had sustained a compression fracture of the fifth and sixth cervical vertebrae. Typically, individuals with a break at that level are able to move their shoulders slightly, and their elbows, but their hands and legs are completely paralyzed. Such was the case with Jatich. He was put into traction for eight weeks in an attempt to stabilize his condition. "Slowly I started getting some feeling back down my arms and elbows. I could move my arms, but not my fingers or my hands," he said.

Two months after his injury, Jatich was transferred to Highland View Hospital in Cleveland. That is where, in January 1978, he met Hunter Peckham, an FES researcher who was looking for someone willing to have experimental electrodes implanted in his hands. Peckham's goal was to reverse atrophy, bring back muscle tone, test and improve electrodes, and eventually produce coordinated movement of paralyzed hands. Jatich was a perfect candidate for Peckham's experiments because he had some shoulder and arm movement—he could raise his arms, but if he tried to extend them over his head, they would fall into his face—and he could flex his left wrist. And he was more than happy to oblige. The plan was to conduct experiments on Jatich for two months. That has turned into a decades-long partnership during which Jatich and Peckham have bonded like brothers.

At the outset, a hypodermic needle was used to inject single-strand platinum electrodes into Jatich's left wrist. Peckham would then manipulate the electrodes under his skin toward the muscles in Jatich's hand that he wanted to animate. To find the correct location, he would periodically shoot spikes of electricity to the electrode to see which muscles twitched. Once he was assured he had arrived at the right place, he removed the hypodermic needle, leaving the electrode in place, with the lead protruding through

Jatich's skin. There were times when as many as ten electrodes extended from the top of his wrist, with an equal number coming out from under it. The external ends of the electrodes were soldered to gold pins, which according to Jatich, "looked like stakes they would stick into Dracula's heart." They were plugged into a junction box strapped to his arm that was attached to a computerized controller by a strand of wires. Each time electrodes were placed into Jatich's hand, Peckham and his team had to wait two weeks for the wounds to heal before they could run tests. Nonetheless, through an exhaustive process of placement, replacement, and testing various stimulation sequences, Peckham and Jatich were able to create a series of different hand grasps enabling Jatich to hold and use various objects, such as eating utensils, an electric razor, and a pen.

But the team of researchers, which now included Jatich as a full-fledged member, was far from having a practical system at hand. The computer power required to control Jatich's electrodes in a coordinated fashion still required a lot of space and was far from portable. This meant that Jatich had to go to Peckham's laboratory to be hooked up to the computer. Additionally, he had no way of issuing commands to the system to let it know what he wanted his hand to do. That was done by the scientists controlling the computer. And the electrodes sticking out of Jatich's wrist were prone to breakage. It was not uncommon for an electrode to break after being in his hand for no more than a week, which meant replacing it and allowing the tissue around it to heal before testing could resume. Even though the broken electrodes were removed, bits remained under Jatich's skin, and to this day they sometimes emerge on their own. "I'll look down and see little shiny things protrude through my skin," he said.

Fortunately for Jatich, and for the entire field of neural prostheses, electronic equipment was shrinking rapidly in the 1980s, which made it possible to eventually fashion a system that Jatich could carry on his wheelchair. But for him to operate the system on his own required that he be able to relay his desired movements to the computer. The solution that Peckham developed took the form of a joystick-like device taped to Jatich's right shoulder, the one opposite the implanted hand. This prevented movement of the hand from inadvertently actuating the system. A separate on/off switch was taped to the center of Jatich's chest. With the limited arm movement he retained, he was able to swing his hand up to hit the switch. The system was capable of creating two types of grasping motions. One strike of the switch readied the system to perform a key-pinch hand grasp, in which the hand makes a fist and the thumb comes down on top of the folded fingers. This configuration is especially good for holding eating utensils. When Jatich hit the switch a second time, the system would go into the palmar grasp mode, in which the hand wraps around an object, such as a glass. Once the system was turned on and the desired mode selected, Jatich could begin to operate his hand by shifting his shoulder to activate the joystick. A forward movement caused his hand to close. A sharp jerk upward would lock the chosen grasp into position. Moving his shoulder downward would unlock the hand, and a backward shoulder motion opened the hand.

When he first rolled out of the laboratory with a portable controller, a joystick, and an on/off switch, Jatich's muscles were animated by four electrodes protruding through his skin, but he was now able to use the system whenever he wished. Meanwhile, Peckham's group was working to develop completely implantable electrodes. By 1986, they had achieved their goal, creating a system they dubbed the Freehand. Jatich, of course, got first dibs at receiving the new system, an opportunity he grabbed at.

During the implant surgery, eight platinum electrodes encased in a soft, inert silicone rubber insulator, were inserted into Jatich's body through a small slit approximately an inch long under his left arm and tunneled down into the appropriate muscles of his hand and forearm. The electrodes were plugged into a connector, also implanted under his arm, from which another set of matching wires were run under his skin to a receiver-stimulator that was slid into a pocket created under the skin on the left side of his chest. To use the system, a caregiver must tape a small transmitter coil onto Jatich's chest directly over the implanted receiver-stimulator. When he commands the system into action by moving his shoulder-actuated joystick, his wheelchair-mounted computer sends command signals to the transmitter—essentially a small radio station that communicates electromagnetically with the implanted device. It sends both control information and power to activate the electrodes through his skin. After receiving the radio frequency signals in analog form, the receiver decodes the information so it can determine which electrodes to stimulate at what current intensity. A portion of the incoming radio signals is converted to electrical energy to provide power to the electrodes.

Complex as this sounds, the process is not unlike early crystal radios that had no batteries and were powered by electromagnetic radio waves. The entire system is programmable to allow for differences in users' range of motion. Once the parameters are set for a specific user, an "expert" software system automatically makes decisions about the coordination between the different muscles that need to be activated for a given grasp.

It took over a decade of human testing and numerous improvements along the way, but in 1997 the Freehand became the first implantable FES device to receive FDA approval. NeuroControl, a company that Peckham helped establish, obtained $4.5 million in venture capital and began to manufacture and market the Freehand worldwide. With the implant, hundreds of quadriplegics gained newfound independence, now able to feed and groom themselves and perform numerous other tasks that they could not do on their own before. But despite a substantial potential market for the device, the Freehand was not a commercial success.

Injuries that damage the nerves running through the spinal cord have rendered some 250,000 people in the United States either paraplegics or quadriplegics, and according to the National Spinal Cord Injury Statistical Center, approximately 11,000 additional spinal cord injuries occur in the United States each year. About half the total number are quadriplegic, the other half paraplegic. It is estimated that half of the quadriplegic population retain enough movement to make use of the Freehand. Yet despite what Peckham called an "overwhelming" acceptance from patients and surgeons, NeuroControl deemed that producing the Freehand was not a viable business and ceased production in 2002. Clearly unhappy over NeuroControl's decision, Peckham is no longer affiliated with the company. And he hastens to point out that the decision to cease Freehand production was economic, not functional. "The devices performed better than we had any right to expect them to," he said.

There were a number of factors for the Freehand's commercial failure, including the substantial expense involved in obtaining FDA approval, questions pertaining to reimbursement for the cost of implantation, and a culture clash between business people wanting to get as many units out the door as possible and clinicians who wanted to proceed more slowly.

Despite this setback, Peckham considers the Freehand venture a success, given the fact that it was the first motor prosthesis approved by the FDA and it is still being widely used. In fact, far from dissuaded, Peckham and his associates at the Cleveland FES Center have since pressed on with efforts to improve the hand control system, believing that the demand is there and that it can be a commercial success if the economics are managed differently. This is a driving force behind Peckham's focus on having scientists at the FES Center create a toolbox of components that can be used in various systems. As a result, the basic technology and some of the same elements, such as electrodes, that were initially developed for the Freehand are now finding their way into stand and step systems to enable wheelchair-bound individuals to rise up and step out on their own.

One facet of Freehand technology that was ripe for improvement was the shoulder-operated joystick. As a patient-control device, it was acceptable for basic open/close commands, but if the system was to provide finer grasp control, a more sophisticated mechanism was needed. And having to tape the joystick and the on/off switch onto the user's chest was certainly undesirable. Additionally, the shoulder-mounted joystick configuration could not be used if bilateral implants were ever to become a reality, since moving one hand could inadvertently activate the other hand. The ideal situation would be to have a control system that could be completely implanted in the body along with the electrodes and the receiver-stimulator. Bypassing the damaged neural highway in the spinal cord by using recording electrodes implanted in the motor cortex of the brain, and sending the signals generated there directly to the muscles, thereby closely mimicking the body's own control system, would be the optimal control system, but that was a dream for the future.

Looking for a more immediate solution, Peckham's team struck upon the idea of using the ability Jatich retained to flex his left wrist as a control signal. Their idea was to place a magnet in one wrist bone and a magnetic sensor in an opposing bone. Wrist movement would change the relationship of magnet to sensor and signal the system to manipulate his hand. A flick of the wrist, they reasoned, would be a lot less awkward than a shrug of the shoulder, and because the wrist can be moved in degrees, it would provide finer control. Pursuing this idea, the Cleveland FES scientists came up with a concept they called the IJAT, for implantable joint angle transducer, which was designed to control ten electrodes rather than the Freehand's original eight. Jatich was again asked if he would like to be the first to receive the IJAT system, but this time he hesitated. His concern was born of the fact that implanting the new system would require removing the old Freehand that he had been carrying around in his left hand for about ten years and had become particularly attached to. To assuage his concern, Peckham offered to implant a second conventional joystick-operated Freehand in Jatich's right hand, giving him bilateral hand control. Nonetheless, Jatich recalled thinking, "Why remove something that's been working all these years? I never even had a broken electrode." In the end, however, he couldn't resist his instinct to be there first and to help advance the field.

So on September 27, 1997, in a particularly tricky operation, surgeons hollowed out small recesses in Jatich's left wrist bones into which they placed the IJAT's magnet and opposing sensor. Their challenge was to carve out recesses in the bone that would allow the two parts of the system to be as close to each other as possible without cutting so deeply into the bone that it would be prone to break. In the end, they did their job well, placing the sensor and magnet a mere 3 millimeters apart, with no residual negative effect on Jatich.

To turn his new system on, Jatich strikes a button on his wheelchair, rather than having to hit a switch on his chest. The wires that feed electricity to his muscles still run in his arm to a receiver-stimulator implanted in his chest, which communicates with a transmitter that has to be taped to his chest—the only external part of the system that must be worn on his body. And instead of the two original grasps, the IJAT is programmed with five, which Jatich scrolls through each time he strikes the on/off button, with the one he uses most being the first to come on. Once the appropriate grasp is selected, he flexes his wrist to begin closing it. As he progressively moves his wrist up, the grasp tightens. When he reaches the optimal amount of pressure, he hits the wheelchair-mounted switch again, and it locks the grasp in place so he doesn't drop whatever he's holding. Another push of the button and the grasp unlocks. Learning how to adjust the proportional control took a while, since Jatich's only feedback is visual and the grasp needed to hold a Styrofoam cup is considerably different from that needed to hold a soda can. But the IJAT has since become such a part of Jatich that when an external component needs repair, he finds it to be traumatic. "When it doesn't work, and I have to have it fixed, and I have to use braces again to eat with, it's like part of my body is gone again," he said.

A few months following the successful implantation of the wrist system, a conventional eight-channel Freehand system was implanted in Jatich's right hand, giving him bilateral hand control. The second system, operated by a joystick mounted on his left shoulder, enabled Jatich, who is right-handed, to wield a pen again without someone first having to strap a brace on his hand for him. The two systems also enabled him to accomplish tasks that most take for granted, such as lifting a bottle with one hand while pouring from it into a glass held with the other or holding a salt shaker in one hand while grasping a fork with the other. With the ability to grasp a pencil in each hand to clank away on his computer keyboard, Jatich was also able to resume his career as a draftsman, which he felt was over when he became paralyzed. One of his first assignments was to do engineering drawings for a test apparatus to measure the wrist strength of paralyzed individuals. All of this, he said, "was a big emotional change in my life."

Successful as the devices implanted in Jatich are, many quadriplegics do not have enough wrist or shoulder motion to operate such systems. To make this technology available to them as well, the FES Center researchers turned their attention to harnessing whatever slight muscular movement a patient may retain anywhere in his or her body to manipulate a hand control device. Their idea was to record the minuscule myoelectric impulses that muscles give off when they are activated and translate them into commands for an FES system. Electrocardiogram (EKG) machines, which record heartbeats, utilize the same principle. Myoelectric activation technology takes advantage of the fact that people who have essentially no movement can, in many cases, flex a muscle in an almost imperceptible way. Even the slightest movement of a facial muscle, for example, gives off a minute current that can be detected by recording electrodes, then amplified and processed by a computer to activate electrodes embedded in muscles.

The resulting system, which is saddled with the ungainly name "implantable stimulator telemeter," or IST-12, incorporates twelve channels, rather than the ten Jatich has in his wrist and the eight in the shoulder-actuated system. The extra electrodes enable the system to record from two muscles simultaneously. One muscle can instruct the user's hand to open and close, while another muscle controls elbow extension, for example.

FES Center researchers received FDA approval to test the IST-12 in twelve quadriplegics. In an unusual turn of events, Jim Jatich was not the first person to receive the system, though he still can claim something of a first with this system; the recording electrodes were initially tested in his platysma—a thin muscle on the front of the neck that connects to the chest and is used when grimacing—to determine their efficacy. With that test successfully completed, the first implantation of a complete IST-12 system took place in June 2003. The recipient was a twentysomething college student. In addition to having electrodes placed in his hand muscles, electrodes were implanted in his trapezius muscle at the base of the neck to aid with shoulder and elbow movement problems resulting from peripheral nerve damage. This kind of damage involves the nerves that are in the muscles themselves, as opposed to those going through the spinal cord, and is not uncommon in people with spinal cord injuries. When such damage occurs, the surrounding muscle atrophies, causing significant weakening. Prior to receiving the IST-12 implant, the young man simply could not use his arms. Within a year of implantation, he was able to feed himself and take some of his own class notes.

The second IST-12 recipient was a woman in her thirties who, ironically, trained helper dogs for the wheelchair bound prior to incurring her own spinal cord injury. She underwent the implant procedure six months after the first recipient and progressed even more rapidly because she did not have peripheral nerve damage. Within several months of surgery, she was able to feed herself, write, and perform functions that most people take for granted but are usually beyond the reach of quadriplegics, such as brushing her hair and applying lipstick.

The long-term goal of the Cleveland researchers is to develop a networked neural prosthesis system that will turn the paralyzed body into the human equivalent of a wired house into which "off-the-shelf" components can be plugged in when the time is right. It can also be thought of as an Internet in a body, with each component being an independent entity that is connected to a larger network, much as computers connect to the Internet. And as with the Internet, each module will exchange information with the other modules in the body over the network. The system will consist of a standard core module that will be implanted in the patient and will contain a battery and a network that will allow various electronic devices to communicate with each other over a central system.

The network system offers multilimb control as well as bladder and bowel function and respiration, as needed. And because the system will be completely implanted, the patient will not have to wear any external equipment. In essence, this will be a manmade nervous system. The core unit would be the brains of the operation, with the backbone cables running from it serving as the system's nerves. Currently, a completely new FES system has to be designed each time a new application is called for. A networked system would use components that have multiple applications, making the design process considerably less onerous, and the system less complex. From a patient's perspective, having a fully implanted system would provide greater independence, since there would be no need for an assistant to attach an external transmitter to his chest, as is required even with the IST-12 myoelectric system. This would make the system available twenty-four hours a day, seven days a week, so a user who wants to scratch her or his nose in the middle of the night, can. And while systems with external components cannot be used in the shower or in a swimming pool, fully implanted systems can be used wherever the wearer goes.

"Can you imagine if I would just get up in the morning and everything is internal? I wouldn't have to put anything on. Someone would just help me get into my chair, and I'd be all set," said Jatich, eagerly anticipating the day when he can be the first to receive a networked system. "I wouldn't have to tape an antenna on, I wouldn't have to make sure the batteries are charged. I'd be able to go into the shower and wash myself with my own system."

In a patient with complex paralysis, a backbone of cables would run from the central core unit down each limb so that various stimulators and sensors can be plugged in as needed. "That will allow us to create on the fly the kind of system needed for any particular patient," said FES Center clinical research director Kevin Kilgore. A case in point would be an individual with a standing system who later becomes eligible to receive a walking system. With current technology, the standing implant is an eight-electrode apparatus and the walking system requires implanting a second eight-channel system. If the patient were implanted with a networked system, upgrading from standing to walking would involve a minimally invasive surgical procedure to implant a second stimulation module. The most significant surgery would involve initially implanting the core unit. Implanting the individual modules for muscular stimulation, or to send impulses to block spasticity, or to record electromyographic signals, or to determine a limb's position in space would require relatively minor surgery. And as each new module is added, the entire system would recognize its presence and establish a priority for that module. If a user were losing his or her balance, for example, righting the person would take processing precedence over monitoring bladder pressure.

"Technically, we know we can do it. We haven't done it yet, but we have major pieces of it working," said Hunter Peckham. This makes it realistic to imagine a day, albeit in the distant future, when a paralyzed individual receiving such a system will be able to move about and function in a manner indistinguishable from an able-bodied person. In the meantime, what has already been accomplished in returning movement to the paralyzed by means of functional electrical stimulation is making a tremendous difference in the lives of those fortunate enough to have received such implants. Jatich told of belonging to a group of spinal cord injury patients that holds monthly meetings. Sometimes the meetings are held in a member's home, in which case virtually all the members show up. On some occasions, the meetings are held in a restaurant and attendance drops significantly. "They didn't want to be seen in a restaurant having someone cut up their food and feed them, so they didn't show up," explained Jatich. He remembered one woman in particular who was extremely withdrawn and reserved. Then she received a Freehand. "She was able to put on makeup, bake cookies and do other activities with her children. She became so outgoing, she joined the spinal cord injury group, went back to school, and goes out to restaurants. She changed completely into a different person. That's the kick I get out of it, how people's lives change."

As for himself, Jatich said simply, "How do you repay someone for giving you the use of your hands back?"

7
Looking Back at
an Empty Wheelchair

J ennifer S. French walked down the aisle at her wedding. There is, of course, nothing unusual about that, unless one considers the fact that she is a quadriplegic. The story of Jennifer's trip to the altar, made possible by the electrodes implanted in her legs, began on March 13, 1998—Friday the Thirteenth, to be exact. The moon was full, and French, then 26 years old and an accomplished snowboarder, was enjoying night boarding at a New England ski area with her boyfriend, and now husband, Tim. It was a beautiful evening for "spring skiing," with the moonlight glistening off the snow, helping illuminate the floodlit trails. During this time of year, it is not unusual for the daytime warmth to melt the snow on the mountains into what skiers sometimes call "mashed potatoes." Then, when the sun goes down, the cold night air freezes the snow on the mountain, turning it into sheets of ice. As French was negotiating an expert trail, she hit just such an ice patch, lost control, and skidded off a ledge, dropping 40 feet or so with her snowboard still firmly attached to her feet. She careened off a number of trees before coming to a stop. In those few seconds, Jennifer sustained a C7 incomplete spinal cord injury. In other words, she instantaneously became a quadriplegic.

The spine is made up of twenty-four separate vertebrae, divided into the cervical, thoracic, and lumbar regions, in respective descending order. The cervical section of the column is made up of the seven highest vertebrae, identified numerically from the neck down. If the nerves that run through the spinal cord are damaged, some level of paralysis and loss of sensation usually occurs below the damaged area. In French's case, the seventh cervical vertebra (C7) was shoved up against her C6 vertebra, which damaged but did not completely sever her nerves at that point. Hence, her injury is referred to as incomplete. Some people with incomplete spinal cord injuries retain some mobility and sensation. Others lose one or the other, while some lose both. French fell into the latter category.

Nonetheless, she said, "I was very fortunate that the ski patrol knew how to handle spinal cord injuries, and that I had methylprednisolone in my system within forty-five minutes of the injury." When the nerves of the spinal cord experience a trauma, the body's immune system deals with the damage through an inflammatory process that, paradoxically, can increase nerve damage. If administered soon after the trauma, methylprednisolone, an anti-inflammatory and immunosuppressant steroid, can minimize the residual damage. The fact that French received the drug relatively soon after her injury may well have helped her in her fight to regain mobility.

Originally from suburban Cleveland, French, a 5-foot, 6-inch, 125-pound bundle of energy, was living in Manchester, New Hampshire, with Tim, an airline pilot, at the time of her accident. They settled there to enjoy the New England outdoor lifestyle. Both had been skiers before taking up snowboarding, and Jennifer considered herself a "pretty good snow-boarder," able to negotiate even the most difficult trails on a mountain. Her love of the sport has led her to refuse to identify the mountain on which her accident occurred, nor has she sued the company that owns the ski resort, because she does not want to give it or the sport a bad name. "It's simply that I love skiing, and I've always been very cognizant of the fact that ski resorts get bad publicity, and that there are some frivolous lawsuits," she said. "It increases the price of everybody's lift pass, so I've always refrained from saying what ski resort it was."

French has a personal interest in doing what she can to keep the price of lift tickets down, since a mere eight months after her injury, she was back on the slopes, this time using a "sit ski"—a chair built low to the ground and mounted on a single ski. Many a talented disabled person can negotiate difficult trails using this device with greater dexterity and speed than many able-bodied skiers. "I still love to ski. It's a wonderful sport," she said. "People need to understand that there's personal risk with any type of sport or anything that you do. I thoroughly believe that."

Always an adventuresome soul, French received a bachelor's degree with a major in aviation from Bridgewater State College, a small institution located in Massachusetts, 28 miles southwest of Boston. She is a licensed private pilot, and during her college days worked part-time as an air traffic controller out of Warwick, Rhode Island, and in operations at Boston's Logan Airport, with an eye toward becoming a professional pilot. Another part-time aviation job while a student, this one at Wiggins Airways, brought Jennifer in contact with Tim, a pilot for the company. He asked her out, but she refused at first, not wanting to mix business with pleasure. During her last day on the job, however, she told Tim she was interested. They began dating, but it wasn't long before Jennifer decided to take advantage of a full scholarship she had been offered to work toward a master's degree in marketing at Wichita State University in Kansas. She and Tim maintained their relationship long distance. When she returned to the Northeast, MBA in hand, Jennifer and Tim moved to New Hampshire, where they lived the nineties lifestyle to the hilt, working between sixty and seventy hours a week, then snowboarding, hiking, or fly-fishing on the weekends. Jennifer's accident drew a drastic, but not permanent halt to those activities.

At the time of her accident, most of the neurosurgeons in the immediate area were out of town attending a convention, so French was flown to the University of Vermont Hospital in Burlington, where she spent a week in the intensive care unit being stabilized. During that stay, her doctors told her that there was a 99.9 percent chance that she was going to live her life out as a quadriplegic in a wheelchair. They "sat me down with my boyfriend and my parents and said this is the way life is going to be: 'You are not going to walk again, you are not going to work again, you are going to be a quadriplegic and in your wheelchair for the rest of your life,'" she recalled. "We were handed a big stack of forms. They said, 'Start your Social Security paperwork.'" French said she understands that the doctors who imparted this information were trying to prepare her for the worst, but she added, "It is a horrible thing to put in somebody's brain." And she, for one, was not going to accept it.

When French spoke of her fight to recover, she never used the singular "I," but rather "we." She explained that spinal cord injury is "a very traumatic experience not only for the person going through it, but for the loved ones around the injured individual as well. Everyone is trying to cope with their emotions."

Doing what mothers do, Jennifer's mom asked her to return to suburban Cleveland so she could care for her daughter. Jennifer and Tim had other ideas but not before an emotional discussion about the future of their relationship. She told him he could leave her at any time and she would not be angry. He wouldn't hear of it. They decided to stay together in New Hampshire. "We decided that we would be a team," said French. Her parents and her employer also became intrinsic members of the team. "We were determined as a family that I was going to be among that 0.1 percent who would walk again. We were going to do everything in our power to make sure that we got the best rehab and find a solution to this. We were going to go out and educate ourselves as to what in the world spinal cord injury is. What is available? Maybe these doctors don't know something that is going on in the world. So we as a family went out and educated ourselves about spinal cord injury and available therapies."

Despite this almost superhuman resolve, French admitted going through frustration, anger, and depression—the usual stages experienced by individuals who are newly paralyzed as they come to terms with their plight. Some push through those emotions relatively quickly, others get stuck. As for French, "I was very fortunate in that I pushed through rather quickly," she said. "Fortitude" is a better word to apply to French than "fortunate," since her ability to fight against all odds is in her nature.

Shortly after being told she would likely never walk or work again, French was transferred from the University of Vermont Hospital to a rehabilitation facility in Concord, New Hampshire. She opted for a facility close to her Manchester home, rather than one of the more well-known, urban rehabilitation centers, because, she said, "I was better off trying to get myself back into the life I knew as quickly as possible." A significant part of that involved getting back to her job as a marketing analyst and project manager with her employer, PC Connection, then a small technology company involved in the direct marketing of computer equipment and software, and now a Fortune 1000 company. Her job included taking on special projects for senior management to help with the growth of the company. French knew that though her body was not functioning properly, her mind was in perfect shape, and she felt that using it would help her body mend. She also knew that her work required that she be able to write and type, so she feverishly set about regaining those skills through physical therapy. She worked in a pool and on a standing machine that got her into an upright position, albeit without using her own muscles. "There were days when I got frustrated, saying, 'Oh, my God, how am I ever going to do this?' but for the most part we were very determined," said French.

Ever so slowly movement and sensation started to return. "We were able to start to move my arms," she said. As arm movement returned, she was fitted with knee, ankle, and foot braces, known as KAFOs, which extended from the bottom of her feet to her hips and allowed her to stand between parallel bars. Within a few months, French was able to operate a computer again, enabling her to go back to work. She set up an office in her room in the rehab center, and it wasn't long before she was receiving more pages from her employer over the public address system than were the house doctors. "It was very therapeutic," she said. The ability to type also freed her up to do battle with insurance companies to obtain approval for additional physical therapy. Railing against the practices of some health care management organizations, French said, "Years ago with a spinal cord injury you would be in the hospital for a year or two. Now when you have a spinal cord injury you go through rehab very quickly, you're out within ninety days. And the insurance companies typically will not pay for more than that. They will send you out on the street whether you're ready or not." A strong believer in "the wheel that squeaks the loudest gets the grease" school of thought, French spent hours using her newfound typing skills to pound out appeals to her insurance companies. Her success at that, coupled with her ability to pay for additional treatment on her own, enabled her to keep at her physical therapy long after most would settle in to life in a wheelchair. Even after she was sent home, French religiously maintained her exercise regimen. "I have a great physical therapist. He would wake me up at six every morning to make sure I'm doing my exercises. I still do his exercise routines in the morning because he really got me into it. You need to get up, and you need to get exercising, and you need to take care of these muscles. Otherwise, you are going to end up with other types of injuries or secondary conditions. He pounded that into my head," said French.

Jennifer can now move not only her hands and arms to some extent but her upper abdominal muscles as well, and some of her tactile sensation has returned. "I can feel light touch and pain. I can feel hot and cold but have some problem with warmth, and there are a few blank spots on my legs where I don't have sensation."

And now Jennifer can stand under the power of her own muscles.

As she fought her way back from virtually total paralysis, French learned of the availability of stationary bicycle devices that enable individuals with paralysis to use their own muscles to power the pedals. Electrodes placed on the surface of the skin, known as surface stimulators, provide the electricity to activate the muscles that would normally be provided by one's own nervous system. The use of such devices helps prevent atrophy of paralyzed muscles while providing a cardiovascular workout as well. Because there was no such device in New Hampshire, French had to travel to neighboring Massachusetts to use one, and she applied to her insurance company to pay for her bicycle therapy sessions. Her request was rejected, but clearly not one to take no for an answer, she fought back. It was while she researched her third or fourth appeal that she came across information about the Cleveland FES Center. Her parents, living near Cleveland, went to the center to learn more about it.

Established in 1991, the Cleveland FES Center focuses on the research and development of technologies that utilize implanted electrical devices to restore movement and various other functions to paralyzed individuals. The center had recently found success with the Freehand. One of center's goals in designing this device was to create a toolbox of FES components that could be used in other neural prosthetic devices as well. When French learned of the center, a group of researchers there was adapting the Freehand technology to a system intended to allow wheelchair-bound people to stand and eventually take steps using their own muscles. The appeal to French was irresistible. With such a system, "I wouldn't have to use my KAFO braces anymore. I would be able to independently stand, and I would be able to exercise these muscles as much as possible," she thought at the time. "It looked like something wonderful."

She learned that several people had already been implanted with an experimental standing system and was determined to be the next test subject. So she sat herself down in front of her computer and began another letter-writing campaign, this one aimed at Ronald J. Triolo, who heads the Lower Extremity Project at the Cleveland FES Center. Triolo is also an associate professor of orthopedics and biomedical engineering at Case Western Reserve University, and a research scientist at the Department of Veterans Affairs.

BORN AND RAISED IN BUCKS COUNTY, PENNSYLVANIA, Triolo was an outstanding high school student with a particular affinity for mathematics and was enthralled with computers, which were just reaching the desktop stage at the time. He was also fortunate enough to have a physics teacher during his senior year who captured his interest by relating many physical principles to human physiology. Not knowing for sure what he wanted to become, Triolo headed for Villanova University to major in electrical engineering. "I knew a core engineering discipline was good training. Like learning Latin and eating Brussels sprouts, it would be good for me," said Triolo. Throughout his undergraduate days, he maintained the idea in the back of his mind that he would try to apply his knowledge to advance some aspect of health care. Becoming a physician was a possibility. Being a high achiever in high school, Triolo remembered frequently hearing, " 'Maybe, you'll be a doctor.' If you couldn't be a priest you could at least be a doctor," he said. But neither priest nor physician meshed well with his temperament, even though health care and altruism were important parts of his motivation. "I didn't want to be in a position where I needed to make quick clinical judgments. I wanted to make a difference in people's lives but wanted to be able to do so methodically in a rigorous way," said Triolo. Biomedical engineering seemed to be the perfect fit, so following graduation from Villanova in 1980, he started his graduate studies at the University of Pennsylvania and a year later transferred to Drexel University, where he received a doctorate in biomedical engineering in 1986.

It was while studying at Drexel that he was introduced to rehabilitation engineering through the university's affiliation with the Moss Rehabilitation Center in Philadelphia. This field perfectly suited his interests, and Triolo selected as his Ph.D. dissertation topic the "myoelectric control of a powered robotic knee joint prosthesis for above-the-knee amputees." This involved using the small myoelectric signals given off by muscle contractions in the stump of an amputated leg to activate the knee.

Shortly after receiving his doctorate, Triolo took a job at the Shriners Hospital for Crippled Children, also in Philadelphia, where a surgeon was starting a functional electrical stimulation research program. His first day on the job, Triolo was sent to Case Western Reserve in Cleveland, which was already well known for its work in the field, to learn as much as he could about FES and return to Philadelphia to help the surgeon set up his program. While at Case, Triolo worked in the motion study laboratory of E. Byron Marsolais, an orthopedist in the forefront of developing implantable standing and walking systems for people with spinal cord injuries. At that point, Marsolais was working with percutaneous systems, in which electrodes were implanted in the patient's body and leads protruding through the skin were connected to a processor that activated the system. Triolo also got to interact with Hunter Peckham, who headed the Neural Control Laboratory, later to become the Cleveland FES Center.

Triolo spent a year in Cleveland before returning to the Shriners Hospital, where he established a program to determine how FES technology could help children with spinal cord injuries and cerebral palsy. In doing so, he took systems developed at Cleveland, including the Freehand grasping device, and adapted them for use in youngsters. In 1994, Peckham invited Triolo to join the faculty at Case and to take a position with the FES Center. Anxious to be where the technology was being developed rather than at the receiving end, Triolo jumped at the opportunity. When he arrived back at Cleveland, he was charged with taking over the work of Marsolais—who had implanted the first standing system in a human in 1991—and turning it into a rigorous clinical trial. He was in the process of doing just that when Jennifer French came knocking on his door.

"I bugged him so much, he had to let me in," said French. Triolo was politic in his agreement. "Jennifer was insistent every step of the way to have us get to the answer she wanted to hear," he said. "She sent us a videotape of her therapy sessions and either called or e-mailed us every week to try to accelerate the process, but there aren't any shortcuts we can take in this research."

Carefully identifying candidates who will benefit most from the system, and who will enable the researchers to learn the most, is a protracted process involving telephone and personal interviews, a physical examination, and a review of all materials by Triolo's team. This accounts in part for the fact that only four people are implanted with standing systems each year. French was hoping to become the seventh of the original thirteen planned experimental implantees. "What we do is still not routine, so it does take a while to gear up to enroll a new candidate, install the system, and of course the rehabilitation after the surgery is pretty intensive," said Triolo.

French would not ordinarily be eligible for the standing program, since candidates are required to be able to use their arms to balance themselves while standing. This limits eligibility to paraplegics, who retain use of their arms. But through sheer willpower and dedication to physical therapy, Jennifer regained enough strength in her shoulders, elbows, and wrists to overcome this objection.

The fact that French was the first woman being considered for a standing prosthesis and that she was of childbearing age was also a matter of concern to the team. Unlike years ago, when a spinal cord injury precluded a woman from having children, young women with spinal cord injuries can now conceive and deliver healthy babies. With Jennifer's candidacy, Triolo's group had to confront the fact that little if anything was known about the effects of electric stimulation on a developing fetus. They searched the literature and consulted with numerous obstetricians and gynecologists. "We wanted to do our homework before we took the plunge with Jennifer," said Triolo. Their diligence turned up nothing to indicate that electrical stimulation would be a major problem for a fetus, so they determined that French's ability to bear children would not eliminate her from consideration. They did decide, however, that if she were to be selected for implantation, they would modify the consent agreement, enumerating the potential risks to childbearing. They also stipulated that she agree not to use the system during pregnancy. Triolo's team also decided that if any woman being considered for implantation said she definitely planned to have children, the electrical leads that usually run from a receiver-stimulator implanted in the abdominal area down to electrodes in the leg muscles would be rerouted under her belly so they do not cross her midline. And if she were to become pregnant, a minor surgical procedure disconnecting the electrode leads from the connector in her abdomen would be performed to allow for expansion as the fetus grows and to protect the components in her body.

With their concerns assuaged, Triolo's group invited French to come to Cleveland for a personal evaluation near the end of 1998 and then again in July of the following year. Two months later, she received the news that she had fought so long and hard to hear. She had been accepted into the program. From there on, things moved much more rapidly. Her surgery was scheduled to take place in November. At the time, implantees had to commit to remain in the Cleveland area for eighteen months following surgery. This allowed for recuperation, physical therapy, and training in the use of the system and gave the research team time to conduct their own studies, since these experimental implants were designed to be a learning tool for them as much as to give mobility to the subjects. In French's case, having her parents in the area made that requirement considerably less onerous than it might have been, yet relocating from New Hampshire disrupted life for her and Tim, who moved to Cleveland to be with her.

As the surgery approached, Jennifer was apprehensive, but certainly not enough to dissuade her. Devoted to nature and the outdoors, she had some qualms about experimental foreign objects being implanted in her body. But an even greater concern was, what if she went through the surgery and the implanted system didn't work? "On the other side," she said, "there was excitement at having the opportunity to break new ground. The benefits seemed to outweigh the risks. If there's no cure for spinal cord injury, what's the next best thing available today? That is the viewpoint that we took. You have to take the philosophy that if people don't participate in research projects, the cure or better rehab is never going to come about. Some people are very apprehensive about participating in research projects, but my point of view is, if I don't participate, how in the world is it ever going to get better? If it's not going to be me, who is going to do it?"

Jennifer's life was too full to allow her much time to worry about her upcoming surgery. By the time she was ready to move to Cleveland, she had gone back to work full-time. Through Northeast Passages, an organization that works with people with disabilities, she was also able to adapt a canoe so she could once again enjoy the lakes and streams of New Hampshire, and she had even returned to fly-fishing. She moved to Cleveland at the end of October 1999 and immediately set up a new office so she could get right back to work after surgery, which took place on November 11, 1999, only weeks after her move.

For seven and a half hours, doctors worked to implant an eight-channel receiver-stimulator, six epimysial electrodes, two intramuscular electrodes, and several connectors in her body. The epimysial electrodes, which are sewn onto the surface of a muscle, are small platinum disks approximately a half-inch in diameter. Except for the metal surface contacting the muscle, they are encased in a synthetic rubberized material. The intramuscular electrodes look like miniature harpoons with small barbs near the tips to anchor them deep within a muscle to get close to nerves embedded there. These electrodes were injected into French's muscles through small incisions in her skin. Stainless steel leads, also sealed in an elastomer sheath, were run just under her skin from each of the electrodes to connectors implanted on each side of her body. Leads from the connectors were routed to the receiver-stimulator, which was inserted into a pocket created under the abdominal subcutaneous fat.

Great pains were taken well before the first patient was implanted with a standing system to determine which muscles should be stimulated to provide the greatest possible strength and stability. Three muscle groups were identified: knee, hip, and trunk extensors. The selected knee extensor is the vastus lateralis (one of the large thigh muscles), which itself is strong enough to raise an individual from sitting to a standing position. Because strong contractions of these muscles are essential for the patient to stand, more time is spent in the surgical placement of these electrodes than the others. The gluteus maximus (in the rear end), is implanted to provide hip extension and is used during the transition to standing. The semimembranosus (a hamstring muscle) is activated to allow the implantee to remain standing, and the erector spinae (a group of muscles that run vertically down the back) are implanted to help maintain proper posture while standing. Before securing the electrodes on Jennifer's muscles, the surgeons used a special stimulator—one of several surgical instruments that had to be designed specifically for such implant surgeries—to stimulate the muscles at the proposed electrode sites to ensure that the desired contractions would take place. Once they were satisfied that they had the optimum reaction, they placed and secured the electrodes.

The hardest part of the entire process for French was a two-day, postoperative anesthetic hangover. "It was equivalent to a tequila hangover from college, but I didn't even get to party ahead of time," she said. During her ten-day hospital stay, it became increasingly apparent that the surgery had gone well and that French's body was accepting the prosthetic components. Then, shortly before leaving the hospital, any doubts she may have had turned to joy when her newly implanted muscles were given a test run. A transmitting coil was placed over the receiver-stimulator embedded in her abdomen, and the activating processor was turned on. It sent command signals and power in the form of radio frequency signals through her skin to the receiver-stimulator, which proceeded to fire off small bursts of electricity to the electrodes embedded in her muscles. And her legs moved. "Oh, my God! I have this wonderful ability suddenly to move my legs," she thought. "That moment was the first time that I was able to see my leg muscles actually contract and move since the injury. All of a sudden my muscles were contracting, and we were able to do it with the help of these electrodes. It was a really great sight," said French.

But her travails were far from over. The next requirement was that she remain sedentary for eight weeks while the tissue healed around the electrodes, encapsulating them in place with scar tissue. She was limited to just four transfers per day from bed to chair and back. Though staying still was not French's cup of tea, it was made more bearable by the fact that she was able to return to work, which involved no more activity than talking on the phone and typing away at her computer. By the end of eight weeks, French was more than ready to begin the next step in her return to mobility, namely, an exercise regimen using the implant to rebuild her long-sedentary leg muscles.

To operate her newly implanted system, she was given a portable lower extremity control unit, or LECU, which she straps around her waist like a fanny pack. It contains the batteries that power the system and a programmable controller about the size of a VHS video cassette. On the front of the LECU are three buttons, a circular green one for "go," a diamond-shaped blue button to scroll through various programs, and a square, red "stop" button. The unit also has a series of lights labeled "stand," "sit," "all" (for all muscles to fire simultaneously), and "back" (for back muscle activation). There is also a "battery low" warning light and another to warn if the external coil is not connected properly. The LECU includes ports for a battery charger and for a computer to program the system.

Much as French wanted to stand as soon as she received her LECU, it was not programmed to allow her to do so. "They didn't put the standing system in because they knew that I would want to try it," she said. Instead, she was given a series of programmed exercises designed to strengthen her muscles and prepare them to bear the weight of her body, something they had not done for some time. Jennifer went at it as though she were training for a marathon. She fired up her stimulator as many as six times per day. She did leg lifts with ever increasing weights around her ankles and would then activate all eight of the electrodes in her legs and back simultaneously. The electrodes turned on and off at five-second intervals, and French did these repetitions for an hour to an hour-and-a-half each day, for the equivalent of a 5-mile run. Seeing a silver lining to every cloud, she pointed out that "I can lie in bed and do the exercises and watch television, while you're going to have to get out of bed, put on your sneakers and go jog 5 miles."

Once the eight-week exercise period was behind her, French returned to the Cleveland FES Center for her first attempt at standing unaided since her accident two years earlier. Tim, who was now her fiancé, and her parents were present for the occasion. Tim had asked Jennifer to marry him shortly after her surgery. She accepted, saying she was determined to walk down the aisle at their wedding.

After placing the transmitter coil over the receiver-stimulator, she readied herself by moving toward the edge of her chair with a walker in front of her. She pressed the "go" button on her controller. There was a three-second pause and then several beeps as the controller began to send electrical current to the eight electrodes implanted in her back and legs, stimulating the muscles in a programmed sequence. The electrical charge was set to increase slowly in intensity so that she rose to a standing position naturally rather than jerking erect. And finally—after months of preparatory exercise, numerous appeals to be implanted, the discomfort of surgery and the subsequent periods of remaining sedentary followed by intense exercise—Jennifer French stood up. She had to grasp the walker in front of her for balance, but she was standing under the power of her own muscles.

"It's just this very wonderful, liberating experience," she said. "You've built up for all these weeks, all these months, to be able to stand, and suddenly you're standing without braces. Without putting on those big clunky things, you just stand with your own muscles out of your wheelchair. It's not your arms that are holding you up; it's your legs that are holding you up." And, she concluded, "there is nothing better than looking back at an empty wheelchair."

During that first stand, she was erect for a minute or two. Through strengthening exercises, she has increased her capability to forty-five minutes.

THE PRIMARY PURPOSE OF THE STANDING SYSTEM is to give individuals with paralysis greater independence. With it, they can transfer between bed, chair, and toilet, negotiate tight spaces too small for a wheelchair, and reach for items on shelves that would otherwise be inaccessible, all without assistance. In addition to saving the cost and backs of caretakers, there is the clear psychological benefit of independence and self-reliance and a substantial medical benefit to being able to stand and exercise one's muscles. Secondary conditions, such as muscle atrophy, osteoporosis, pressure sores, and urinary tract infections are rampant in people with spinal cord injuries. French, who uses her system to stand for an hour or two a day and to exercise for two to three additional hours per day, has not experienced any such problems.

As for drawbacks, French does see a few. "At the end of the day, you still have a foreign object in your body, and you need to take precautions in terms of infections. It's just like somebody who has a pacemaker and has to take antibiotics before going to the dentist," she said. One implantee did, in fact, opt to have his system removed after developing an infection of the type that is not uncommon among people who receive conventional implants, such as hips. Though the infection could have been treated with intravenous antibiotics, he chose to have the system removed. French had to undergo surgery once following the initial implant procedure to have some electrodes replaced. She viewed that minor setback philosophically, saying, "That's simply part of the risk of being in this research project. It is still research, and it is not a perfect system."

Since receiving FDA approval for the original thirteen experimental standing implants, of which French was one, Triolo has gotten the go-ahead to implant thirteen additional patients, which will take several more years. To keep tabs on the progress of the implant recipients after they leave the Cleveland area, their LECUs are programmed to track their use. Every three months, the units are exchanged for fresh ones, and through the stored data, Triolo's team can determine who has been exercising and standing and who hasn't. French, of course, is at the head of the class. "She is the best possible result," said Triolo.

Surprisingly, most recipients do not use their implants as much as one might assume. On average, most of the test subjects utilize their implants 60 percent of the days they've had them. They use them to exercise about three times per week for about an hour at a time, and they stand once or twice a week on average. But even if an individual does not use the system much, having the option seems to make all the difference. When queried about his disuse, one implantee, who does not use his system at all, made a point of asking that it not be removed.

On the opposite side of the ledger, French doesn't wait for Triolo's crew to contact her. Instead, she gets in touch with them on a frequent basis, asking if they can reprogram her control unit to accommodate new ways she has found to use the system. Although her system was designed for standing and not walking, she uses it to walk by grasping a walker for balance and swinging her legs forward. "My legs stay straight. I put the walker forward, lift, and swing my legs through," she explained. It was with this swing gait that Jennifer walked to the altar at her October 2001 wedding. The ceremony was held in Florida, where she and Tim had moved earlier in the year, in part because of the realization that navigating a wheelchair through New Hampshire's snowy winters is no easy task. They also moved because Jennifer learned of a Paralympics sailing team that has a wheelchair-accessible facility in St. Petersburg. Having enjoyed sailing before her accident, she decided she wanted to race sailboats, which she now does standing at the helm. This does not make the research group in Cleveland happy, since she is not supposed to get the external control unit wet. But they learned a long time ago that attempting to keep French down is a losing proposition. As she put it, "I don't want to sit home and look out the window. I'd rather be out there."

Part of being out there includes doing all she can to help others in her position. In 2003, French co-founded the Society to Increase Mobility, Inc., an advocacy group for persons with paralysis, which has since been renamed the Neurotech Network. The organization had its inception when French attended a neural prosthetic conference sponsored by the NIH and met a hand grasp implantee, a woman with a cochlear implant, a woman with Parkinson's disease who had a deep brainstem implant, and an editor of a neural prosthetic publication. They came to the conclusion that there was a gross lack of information among people with disabilities regarding neural prosthetic devices and that an organization was needed to enable those who have received such devices and those considering them to exchange information. "Our main mission," French said, "is to help both users and potential users obtain information, as well as to provide feedback to the industry, because users are the ones who will help the industry grow and help with the development of these devices."

In the meantime, feedback that the Cleveland FES Center receives from French and her fellow standing implantees is being used to direct the development of the next generation of implants. The researchers have learned, not surprisingly, that smaller and lighter users can stand longer and rely less on their arms for support. One of the challenges is to enable all implantees, regardless of size, to realize the same benefits. Utilizing the brute force approach by implanting more electrodes in more muscles to help with the heavy lifting is one tack the Cleveland investigators are taking to deal with this issue. But doing so requires a new processor to actuate and coordinate the firing of the additional electrodes. Another, more elegant approach involves the development of a new type of electrode that would enable more precise activation of the muscles necessary for standing.

The muscle-based electrodes that French has are placed on and in the muscles near the nerves to be activated. Current flows from the electrodes through the muscles to the nerves, which in turn provide the stimulus to the muscles. The nerves are induced to contract muscles, rather than the electrodes themselves, because doing so requires less current than would be needed for the electrodes to activate the muscle directly. And the less current required, the less chance there is of tissue damage, and the longer the life of the controller's batteries. It follows that the closer an electrode is placed to a nerve fiber within the muscle, the better. But with FES that is not necessarily so. This is because as a muscle contracts, the electrode moves with it. If the electrode is too close to the nerve fiber, the muscle's response to the injected charge may be unpredictable as the nerve moves closer to and farther away from the electrode. FES electrodes have to be placed close but not too close.

To get away from such complexity and to more finely control muscle activation, Triolo's group, as well as investigators at other facilities, are developing nerve cuff electrodes. As their name implies, these electrodes encircle bundles of nerves before they enter the muscles. In vertebrates, the spine is like the trunk of a tree and the nerves are like branches that send off twigs, many of which may supply a single muscle. The principle behind the nerve cuff electrode is to encircle a nerve bundle higher up on the organizational tree than the individual muscle. This permits the activation of individual nerve fibers in the bundle before they branch off to enter muscles, which more closely mimics the body's normal stimulation route and therefore provides more efficient muscle stimulation. This configuration also facilitates the activation of many muscles by a single electrode. By providing finer control of the muscles, this approach could even allow a stepping feature to be incorporated into a standing system without the need for a large number of additional electrodes.

The first generation of FES nerve cuff electrodes requires that the contacts inside the cuff be placed directly on the specific nerve fibers to be stimulated. Subsequent versions will allow the electrode to be electronically tuned after implantation. Each contact will be turned on individually after surgery to determine which muscle it activates and will then be assigned a processor channel. This is similar to the manner in which cochlear implants are tuned after surgery, with each individual electrode assigned a frequency range.

The first humans to receive the experimental nerve cuff electrodes at the Cleveland FES Center will be the heavier standing system implantees who already have the eight-channel system, in the hope that it will enable them to stand as long as the smaller patients. Once the standing system is optimized, Triolo hopes to incorporate nerve cuff electrodes into a sixteen-channel system that will activate more natural standing and stepping.

With the additional firepower of sixteen channels, Triolo and his team may be able to include balance in future standing and stepping systems. The additional channels would make it possible to create a closed-loop system that responds to the user's posture to maintain balance as an able-bodied person's vestibular, visual, and proprioceptor systems do. The vestibular system, located in the inner ear, senses acceleration of one's head and uses that information to drive the leg muscles to maintain balance. The proprioceptor system keeps track of the position of the joints and the pressure on the feet. In individuals who are paralyzed, these systems are operational, but information cannot get from the muscles to the brain and back again.

The initial approach to a balance system is to create an artificial vestibular system that will give a user the ability to fix his or her position in preparation for a change in posture. Normally, when a person knows he or she is going to move in some way, to shake someone's hand or to lift an object, for example, the muscles in the lower extremities as well as in the hand and arm are automatically activated to prepare for the action. FES users cannot do that. In fact, when standing, their implants fight any attempts to lean in specific directions, since they are designed to keep them centered. It is like trying to lean forward while being completely rigid. The plan is to equip the systems with a joystick that will enable users to modulate the flow of electricity to their stimulated muscles so they can lean in whatever direction they wish. This will require that muscles in addition to those stimulated by the existing eight-channel standing system be used, including the muscles that control the ankle, those that press down against the floor, those that lift the toes up, and those that pull the hips from side to side.

The next step in balance system development would be a fully automatic system that uses a gyroscope to maintain a patient's balance in the event of unexpected disturbances, like being bumped. A major challenge to designing such a system is the complexity of taking acceleration readings from the gyroscope in any of three directions (forward, backward, and sideways) and using them to determine which muscles to stimulate, essentially parsing out the small set of sensory inputs into an almost infinite number of combinations of muscular stimulation. Millions of years of evolution have led to an exquisite automatic system that enables our brains to do the reaction computing without so much as a fleeting conscious thought. To try to mimic the natural system, researchers are exploring the use of artificial neural network software to map the route between the gyroscope's output and the desired stimulus. Ultimately, it may be possible to incorporate an automated vestibular system into a standing and stepping system so that someone who is paralyzed can stand and walk without the aid of a supportive device like a walker, though such a system is many years away. In the meantime, as Jennifer French said, "there is nothing better than looking back at an empty wheelchair," with or without a walker.

8
The Dirty Little Secret

"One of the things about spinal cord injury is that everybody sees the outside, as far as you've lost the ability to use your legs, but they don't realize—and it's really demoralizing—that you've lost the ability to go to the bathroom like a normal person, or to even have any feeling to know that you have to go to the bathroom, and having accidents in your pants all the time. It's something you don't talk about except with other people who are in similar situations."

Speaking the unspoken was Holly S. Koester, of Walton Mills, Ohio, near Cleveland, who was a 29-year-old captain in the U.S. Army when in 1990 an SUV she was driving at Redstone Arsenal in Alabama, flipped over and broke her back. Her T7 (thoracic) vertebra was crushed, leaving her without feeling or function "from about the bra line down," as she put it. Since then, Koester has been implanted with a Vocare bowel and bladder neural prosthetic system that has helped prevent the kinds of embarrassing and unhealthy circumstances she described.

Prior to her accident, Koester had been a paratrooper and company commander with the 101st Airborne Division at Fort Campbell, Kentucky. She had been transferred to Redstone but was about to return to Fort Campbell as preparations were underway for Desert Storm in Iraq when tragedy struck.

Always an active person, Koester had run in several 10-kilometer races as well as a half marathon while in the Army. She also played softball and was on the All Army volleyball team. While a student at Fredonia College in New York, she played intramural basketball and soccer. She had a Reserve Officers' Training Corps scholarship to help her through college, and immediately following her 1981 graduation, with a degree in political science, was commissioned in the Army. She was well on the way to making the Army her career when her back was broken. Though she was rendered a paraplegic, Holly Koester did not take her fate sitting down.

Shortly after her accident, she was transferred to the Cleveland Veterans Administration Hospital—the closest veterans' facility to her hometown of Buffalo, New York, with a spinal cord injury unit. Shortly thereafter, Koester volunteered to be a test subject for an experimental standing and stepping system being developed by the Cleveland FES Center, a consortium that includes the VA hospital. But the experience did not work out well. Koester was implanted with electrodes that protruded through her skin, which caused frequent infections. The coup de grace came when it was discovered that she had sustained a broken hip, apparently in the accident that paralyzed her, which threw off her gait when she tried to walk with the implanted system. As a result, her ligaments became stretched, causing her hip to pop out of its socket. An attempt at corrective surgery failed. The result, according to Koester, was that, "I wasn't any use in the walking FES program." But her enthusiasm for the potential benefits of neural prostheses was undiminished.

And then she met Graham Creasey, an English physician specializing in spinal cord injuries. Creasey had recently moved to the United States and was working to have an implantable bladder and bowel device approved by the FDA. The system had been available in Europe for several years, where it was known as the FineTech Brindley Bladder Control System.

Creasey had grown up in Northern Rhodesia, now Zambia. His parents were from London but had moved to Africa, where his father, an engineer, served as a missionary. When Graham was 18 years old, he reversed his parents' footsteps and returned to Britain, where he attended college and then medical school in Edinburgh. During his studies, Creasey found that in addition to medicine, he was fascinated by things technical, which he traces to his father's profession. "My father was an engineer who would like to have been a doctor, and I'm a doctor who rather likes engineering," he said. As a result, he gravitated toward the hands-on field of general surgery. He began to focus on spinal cord injuries when in 1978 he returned to live with his family, who had moved to Zimbabwe, then in the middle of a guerrilla war. He joined the staff of a government hospital and quickly gained experience working with patients suffering from traumatic war injuries. In particular, he said, "I found working with people with spinal injuries a very interesting area. They are mostly young, mentally normal, highly motivated people who had a sudden injury, but were stable and had a long life expectancy."

Another draw for Creasey was the fact that spinal cord injuries leave a lot of room for the application of technology to provide appliances for people to control their environment and their own bodies. This was the technological connection Creasey was looking for. "Sometimes technology gets a bad name," he said, "but there are ways to use it to humanize people's situations, and finding those appropriate ways is really very exciting." He felt that excitement when he learned of a neural prosthetic bladder system invented by fellow Englishman Giles Brindley, who is regarded by many as the grandfather of neural prostheses (see chapter 4).

Brindley had made quite a splash when he implanted electrodes on the visual cortex of a blind woman, but Creasey was more interested in a less dramatic, less complex, but more immediately practical invention of Brindley's that was applicable to people with spinal cord injuries. It was a basic three-channel system that restores bladder, bowel, and sometimes sexual function to the paralyzed.

As recently as the early 1970s, when Brindley began work in earnest on the bladder and bowel system, spinal cord injury patients were frequently hospitalized, and many died of renal failure because they weren't able to properly empty their bladders. Though less so now, improper voiding remains a major problem for paralyzed individuals. This is because they have little or no sensation to alert them to the fact that they need to empty their bladders, and even if they know their bladders are full, they cannot control the muscles that produce urination. Catheterization is an option, but it is inconvenient, can cause infections, and does not always enable patients to void completely.

Brindley thought he could bring his neural prosthetic prowess to bear on this problem by implanting electrodes on the nerves that activate the bladder muscles near where they exit the spine. What he didn't know, however, was whether these spinal roots would survive long-term implantation, so he tested his idea on several baboons for over three years before attempting to implant the system in a human patient. The animal testing also enabled him to determine precisely which nerves he had to activate to stimulate urination, using what he called the "try it and see" approach. He found that what is known as the S2 bundle of nerves coming from the sacral section of the spinal cord innervates the bladder in the baboon. In humans it is usually the nearby S3 spinal root.

Eventually satisfied that the device was safe, Brindley performed the first human implant in 1976, on a woman who had been disabled by multiple sclerosis. The system she received was essentially the same configuration as that implanted in Koester sixteen years later. It consists of three electrodes placed on the spinal root and wired to a pacemaker-like receiver-stimulator embedded under the skin, usually in the abdomen. The user has an external coil that she places over the implanted unit. Radio waves transmitted by the coil through the skin to the receiver-stimulator provide both power and the signals that activate the electrodes. Since most paralyzed patients cannot determine when their bladders are full, they stimulate by the clock, usually four or five times a day, to empty their bladders.

Though Brindley's first patient was able to use the system successfully to induce urination, she retained sensation in her lower body and found the stimulation painful. The next two test subjects, who were implanted in 1978, were both men who had no lower body sensation and were able to use the system with no discomfort. The following year, four more patients received the bladder system, and for the next several years, patients were implanted at a rate of about two per year. After a total of fifteen patients had received the bladder system, the rate of implantation began to increase rapidly enough to justify commercial production, which was undertaken in 1982 by FineTech (now FineTech Medical Ltd.) in Britain. It took a number of years of work for

Creasey and his colleagues at the Cleveland FES Center, but in 1998 the system received FDA approval for clinical use in the United States, where it is now sold under the Vocare name.

In addition to facilitating urination, which was Brindley's primary goal, the system has two peripheral benefits, one being that it can help patients control their bowel movements. People with spinal cord injuries usually cannot control their bowels for the same reasons they cannot urinate normally. And while the inability to void urine can be life threatening, not being able to control one's bowel movements is more of a nuisance, albeit a significant one, than a life-or-death problem. As a result, "once the implant worked for the bladder, we soon thought we'll try it for the bowel, and it worked for the bowel as well, and is for many patients, a great convenience," said Brindley. Another adjunct effect of the bladder system is that it can produce erections in men—though with the advent of chemical means of inducing erections, such as Viagra, most male implantees do not use the Brindley system for that purpose.

The bladder system can offer these additional benefits because the nerves that control the bowel and sexual stimulation are intermingled with those involved in urination. As a rule, some nerve fibers from the S3 and S4 nerves go to the bowel and some fibers go to the bladder, so those two nerves are stimulated to produce urination and defecation. Typically, erection is caused by stimulating the S2 sacral root. Usually implantees use the system to empty their bowel and bladder at the same time, but the controller can be set to facilitate one or the other, thanks to the happy coincidence that the bowel and bladder each contract at different speeds. If the user wishes only to urinate, the controller sends brief pulses of electricity to the nerves, which will not affect the bowel, since it contracts at a considerably slower pace. Conversely, longer, slower pulses are used to initiate a bowel movement without urination.

Creasey first learned of the Brindley system when he met Brindley in 1982, after returning to England from Africa. He was working in the spinal injury unit of a hospital in Edinburgh when Brindley came calling to teach physicians how to implant his new bladder system. Creasey took an immediate liking to Brindley and was fascinated by the bladder system, since it perfectly suited his abiding interest in medicine and technology. During the same period, Creasey learned of other neural prosthetic work being done by researchers at Case Western Reserve University. To learn more, he attended a meeting in Cleveland in 1983, where he met Hunter Peckham, a leader in the development of functional electrical stimulation systems who was to become the founder of the Cleveland FES Center. Because of their mutual interests, Creasey was invited back to spend more time comparing notes with investigators there. "So I started visiting Cleveland, initially for three days, and then three weeks, and then three months," said Creasey. At the same time, he began working on his own to improve the Brindley bladder system.

In 1990, while back in England, Creasey received a telephone call from Tom Mortimer, another Cleveland-based FES pioneer, who was also striving to improve the bladder system. Mortimer wanted to discuss his progress with Creasey.

When a person urinates, the bladder contracts and the sphincter relaxes to allow a steady stream. Following spinal injury, the functions of the bladder and the sphincter are no longer well coordinated, with the result that the bladder and sphincter can contract simultaneously and end up fighting each other. If the bladder wins the battle, incontinence occurs. If the sphincter wins, urine backs up to the kidney, which can lead to dangerous spikes in blood pressure and renal damage.

When Brindley invented the bladder system, he had to grapple with this problem because merely stimulating the appropriate nerves could lead to the same battle between the bladder and sphincter. He cleverly resolved this dilemma by taking advantage of the natural difference in response times of the bladder and sphincter muscles. The bladder muscle relaxes and contracts slowly, whereas the sphincter muscle does so rapidly. Brindley uses this difference to prevent simultaneous contraction by stimulating the muscles in a series of bursts. Because the bladder is slow to respond, it does not relax between bursts, allowing pressure to build. The sphincter, on the other hand, immediately relaxes between bursts, allowing urine to pass. The solution works but is not ideal, because the patient urinates in bursts as well, rather than in a steady stream, and this can result in incomplete voiding.

Creasey and Mortimer were each trying to develop a system that would activate more normal, steady-stream urination. During a telephone conversation, Mortimer told Creasey that he had found a way to relax the sphincter but was having trouble making the bladder contract. Creasey told Mortimer that he could make the bladder contract but couldn't relax the sphincter. They felt that by merging their efforts they might come up with the perfect solution. Mortimer was able to procure some funding to work on the problem, and he invited Creasey to join him in Cleveland so they could combine forces. Creasey obliged, and their work was going well but was far from complete, when at the end of three months, their funding ran out. Creasey made ready to return to England. Then, just a few days before he was to leave, Mortimer attended a party at which he told an acquaintance of the situation. As luck would have it, the person to whom Mortimer spoke was an official of the Paralyzed Veterans of America who said he might be able to help. In short order, Mortimer had additional funding, and he asked Creasey to stay on for an additional two months. "At the end of the two months, he said, 'This is going well, could you stay another three?' " recalled Creasey. "And I'm still here," he said, fourteen years later, although he is not working on the same project.

Creasey and Mortimer spent about three years addressing the bladder-sphincter coordination problem before they decided that Brindley was correct when several years earlier he told Creasey he thought the problem could be overcome but he wasn't sure it was necessary to do so. Creasey said he and Mortimer eventually concluded that Brindley's system worked "pretty well as it was, and perhaps the priority was to get it into more people rather then to spend more time refining it. So I went back into clinical work in spinal units in Cleveland." Creasey also set about the detailed business of procuring FDA approval for the Brindley system, which required implanting human research subjects. Holly Koester was one of them.

She was a perfect candidate for a bladder prosthesis. The fact that she casually referred to urinary tract infections as UTIs, just as most people use everyday acronyms, such as ATM, spoke clearly to the fact that the infections were a regular part of her life. "It's real hard to do a clean self-catheterization, especially when you're brand new at it," explained Koester. "You're sitting on the toilet and if you make a mistake and get your catheter dirty, if there's nobody there to clean it off, you do the best you can." The frequent result is a UTI. So when she learned that Creasey was looking for patient volunteers, she was more than willing. She was also pleased that the Vocare system would help her with what she called her bowel care, since having a bowel movement can require hours on the toilet, and embarrassing accidents are all too common.

Koester tells of giving herself suppositories between 8 and 11 at night, which would occasionally take effect after she got into bed and other times not until she had gotten into her car to make a morning class at Cleveland State, where she was working to obtain a substitute teacher's certificate. Other times, she would be in class and have a bowel movement.

"It's just demoralizing to smell something and not know if you passed gas or had an accident or somebody around you can smell it," said Koester. Then there are the trials and tribulations of trying to clean one's self in a stall in a bathroom where there is too little space. She tells of having to carry around wipes and extra clothes. "I've ended up throwing out a lot of clothes," she said. And she has had to contend with soiling the seat of her wheelchair. "Then you have to take your seat cover off. Or if it soaked through, you are stuck. You use paper towels or whatever in order to try to make it through the day." To Koester, Vocare represented freedom from all that. But there was a drawback.

The timing of stimulation developed by Brindley successfully induces urination, but it does not prevent incontinence. As a result, when people who received the implant went without catheterization, incontinence remained a problem because of spasticity in the bladder and sphincter when the prosthetic was not turned on. The resolution developed by urologists is to sever the sensory nerves that run from the bladder to the spinal cord in a procedure known as a rhizotomy, which is typically performed when a Vocare system is implanted. The downside to this procedure is that it totally eliminates any sensation that patients with incomplete spinal injuries may retain in that area of their bodies. As a result, many patients, especially men, are unwilling to submit to the procedure. This is true even of some patients with no sensation in their lower bodies, who hold hope that someday a repair for spinal cord injury will be developed and are concerned that severing the sensory nerves will make them ineligible for such a cure. This accounts, in part, for the fact that despite its many benefits, the Brindley system has been implanted in only a few thousand people over the years.

Koester, however, was unconcerned. "I kept thinking, 'If they come up with a way that they can connect nerves in the spinal cord, then they're going to be able to connect the others too,'" she said. So in August 1992, she underwent a Vocare implant procedure, which is not a simple one.

Koester was placed facedown on an operating table, and a 4-inch incision was made in the middle of her back. Some bone was removed to gain access to the appropriate nerves. Once located, the nerves were electrically stimulated with a small probe. After catheterization, pressure in her bladder was measured during stimulation to make sure the nerves giving the greatest rise in bladder pressure were selected. Although S3 and S4 usually activate bladder and bowel function, occasionally it is found that S2 and S3 produce those functions. Once the appropriate nerves were isolated, electrodes were secured adjacent to the nerves by small straps of silicone and Dacron. Wires leading from the electrodes were then snaked under Koester's skin halfway toward her stomach. While she was still on her back, the rhizotomy was also performed. Because the motor part of the nerves must be left intact so that the prosthetic device can cause them to activate the bladder and bowel, severing the sensory portion of the nerves is a very delicate procedure, requiring a considerable amount of time. Once completed, the incision in her back was closed and Koester was turned onto her back. Another incision was made in her abdomen so the receiver-stimulator could be placed under her skin. The wires from the electrodes were then run the rest of the way to her abdomen and plugged into the receiver, and the incision was closed. In all, the procedure lasted roughly six hours.

After spending about a week in the hospital, "they told me I was good to go," said Koester. She was back in class within a few weeks.

As it turned out, the bladder and bowel system did not work as advertised for Koester. It seems that as a result of urination problems following her injury, but before she was implanted, her bladder was stretched to such an extent that even with stimulation she cannot void completely. As a result, she must still catheterize herself. She is delighted with the system nonetheless, since it has greatly reduced the incidence of UTIs and allows her to regularly schedule her bowel care. "Emotionally, I'm 100 percent better. It changed my life tremendously," she said.

Though she cannot use it to void completely, activating the implant to induce urination cleans her urinary tract so that she can do what she calls a "clean catheterization." The result is that she now has three or four UTIs per year, as opposed to approximately one a month before receiving the system.

And instead of using suppositories and not knowing when they will take effect, she sits on the toilet and turns on the stimulator, which moves the stool into position for evacuation. "I can take care of everything on the toilet. I have control over timing now," she said. This has clearly increased her confidence. "I'm a substitute school teacher," said Koester. "You can imagine getting a call at 5:30 in the morning asking if you can teach, and having to say, 'No, I can't because I'm doing bowel care'? Now I can take care of everything on the toilet and I don't have any mess to clean up or worry about having an accident while teaching classes."

Successful as the Vocare system has been, Creasey and others are working to make the system palatable to more patients by doing away with the need to cut the sensory nerves to prevent incontinence. The path being investigated involves more closely mimicking the body's normal functioning by stimulating the sensory nerves (instead of the motor nerves) rather than cutting them. The hope is that doing so will send a more normal signal to the spine, which in turn will send a reflex reaction back to the bladder and bowel. If this configuration works, it may well produce a steady urination stream as well, since it is likely that the spinal cord will produce more natural coordination of the sphincter and the bladder. Work is also proceeding on developing a system that will provide sensory feedback to the Vocare user to indicate when his or her bladder is full, so that rather than voiding by the clock, the implantees can do so more naturally, when the bladder is full.

In the meantime, the Vocare implant has enabled Koester to return to the active lifestyle she enjoyed before her injury. She has become an accomplished wheelchair marathoner, having competed in over fifty marathons in seventeen states, with some finishes high in the rankings. Her goal is to do at least one marathon in each of the fifty states. She also participates in various field events at the National Veterans Wheelchair Games, where she took gold in a 2005 air gun event, and she competes at pool and bowling on able-bodied teams.

9
Sound in
the Brain

Marilyn Davidson lives in a neat, unassuming garden apartment complex in Anaheim, California, not far from Disneyland. At the ground-floor front door to her apartment are two doorbells, one of which has a sign above it reading, "Please push this bell." When one does, there is no sound, but through a front window a visitor can see a light flash inside. Davidson, who was born in 1932, walks briskly to answer the door, despite the fact that her sense of balance is nonexistent. The semicircular canals and vestibular portions of her inner ears, which provide equilibrium, no longer function.

Davidson's physical problems came on without warning one warm, sunny day while she was still in her twenties. She and a friend decided to spend the day at the beach. While there, she began to feel poorly. Her stomach was slightly upset and she had an overall "weird" feeling. Assuming she had gotten a bit too much sun, she told her friend she was ready to go home. When she got there, she showered and went right to bed. The next morning, her friend called to see how she was feeling, but Davidson did not answer the telephone. After several attempts to reach her, Davidson's friend called her sister, who lived nearby. She hurried to her sister's house to find that Marilyn was still feeling disoriented and that her hearing was greatly diminished. She rushed Marilyn to the family doctor, who was stumped by her symptoms. Two weeks later, Davidson's hearing returned to normal in her left ear, but she still could hear nothing with her right. "I knew it was more than just a hearing problem. Because I was dizzy, I had a head noise, and my balance was off, but nobody could find an answer," recalled Davidson, who is originally from Nebraska but has spent most of her life in California.

Over the next five years, Davidson went to a number of doctors trying to discover what was wrong with her, all to no avail. She eventually decided to just get a hearing aid and give up her search. But when she saw an audiologist to be fitted with a hearing aid, he told her that the results of the tests he conducted on her were the strangest he had ever seen. He suggested she make an appointment with William House, an ear, nose, and throat specialist, who along with his brother Howard, specialized in ear surgery.

Davidson took the audiologist's advice and set up an appointment. The first time House met Davidson, he told her that he suspected she had a brain tumor, but he wanted to conduct some tests to confirm his diagnosis, and the test would require that she be hospitalized. "They took fluid out of my spine. It was a horrible test," said Davidson. "Dr. House came in the next day and said, 'You have a brain tumor and it has to come out.'" That was in 1964.

The tumor, which was located near Davidson's auditory nerve, was successfully removed, but in the process the auditory nerve had to be severed, completely eliminating what little hearing she had in her right ear. Yet for ten years following the surgery, Davidson had no further symptoms. During that period, she moved to Colorado and was seeing another physician. Then, the hearing in her left ear, the good one, started to diminish to the point that she had trouble hearing the telephone ring. Her local doctor told her he suspected another tumor, but she refused to listen. Instead, she stopped going to him. "I had a cold. That's what I kept telling myself. It's going to get better," said Davidson.

But it didn't get better. Eventually, the obvious could not be denied, so she called William House and made an appointment. When she got to the Otologic Medical Group, a battery of tests were immediately begun, but the diagnosis was so obvious they were never completed. Following the first test, Davidson was on her way to the X-ray department when she was stopped in the stairwell and told she didn't need the X-rays, that Dr. House would see her immediately. As she entered his office, she blurted out the inevitable: "Dr. House, I have another tumor don't I?"

"Yes, you have. You have another one on the right and one on the left," she recalled him saying. But the news was not all bad. House told her that he would be able to perform a procedure on her left side that would not remove the tumor but would take the pressure it was causing off her auditory nerve so she could retain some ability to hear. The procedure was successfully performed in 1976. But the tumor continued to grow, and three years later, it had to come out. Davidson had no option. It was either removal of the tumor and total deafness or the significant possibility of eventual death from the disease.

The diagnosis was neurofibromatosis type 2, or NF2. It is a rare but terrible hereditary disease caused by a single gene mutation that occurs in one of every 40,000 live births and affects some 3,000 to 4,000 people in the United States. The defining characteristic of the disease is that it causes benign tumors to grow on the vestibular portion of the eighth cranial nerve— the nerve that serves the vestibular and cochlear functions. The facial nerve is very close by and is frequently affected by these tumors, causing paralysis on one side of the face. As a result of this paralysis, patients often cannot blink, which leads to eye problems, and the lips and tongue can be affected as well. In addition, the disease can cause tumors to grow on the spinal cord, and although they are benign, they cause a lot of havoc by pushing against or pinching off nerves. NF2 used to be called the "wasting disease" because in advanced stages of the worst cases, paralysis causes patients to look like concentration camp skeletons. Some patients die of pneumonia or asphyxiation as a result of tumors interfering with breathing. Davidson's mother died at age 56 from NF2 but not before the disease both deafened and blinded her, and another sister died on the operating table at age 19 while having an NF2 tumor removed.

The disease affects the insulation that surrounds the neurons of the peripheral nerves supplying the muscles. The insulating material, called myelin, serves the same purpose as rubber insulation around electric wires. This insulating myelin is produced by two different cell types—Schwann cells in the peripheral nervous system and oligodendrocytes, a type of glial cell, in the central nervous system. As the nerves travel from the periphery to deep in the brain, the Schwann cells hand over the job of nerve insulation to the oligodendrocytes. It is at this junction that genetic damage in NF2 causes Schwann cells to replicate out of control and become growths called schwannomas. It is the schwannomas that do the damage.

The growth of NF2 tumors on the vestibular nerves was frequently considered a virtual death sentence because once they grew so large that they had to be removed, the mortality rate from the surgery was approximately 40 percent. Then, in the 1960s, William House developed a new surgical method of removing the tumors known as the translabyrinthine approach, which dropped the mortality rate to virtually zero. But while this procedure saves lives, it does so at a high price. Because the tumors are almost always integrated with the auditory nerve, which runs right near the vestibular nerve, both have to be severed during surgery, leaving those who survive completely deaf on the side on which the surgery is performed, and with an impaired sense of balance.

Though the disease inevitably causes tumors on both of the vestibular nerves, the tumors do not necessarily grow at the same time and the same rate. As a result, a patient may have surgery on one side but retain hearing in the other ear until the growth on the second side becomes large enough to require removal as well, as was the case with Marilyn Davidson. About fifty to seventy-five individuals undergo this tumor removal surgery in the United States each year. Since severing the auditory nerve completely breaks the connection between the ear and the brain, cochlear implants do these patients no good.

When to perform this tumor removal surgery is a difficult decision. With genetic testing and ever improving imaging technology, tumors can sometimes be caught and removed early enough that the patient's auditory nerve does not have to be sacrificed. With the patient's hearing preserved on one side, surgeons can then be more aggressive in going after tumors on the opposite side, should surgery be required. If hearing has to be sacrificed when the first tumor is operated on, however, the tendency is to allow the inevitable tumor on the opposite nerves to continue to grow until removal becomes absolutely necessary, thereby preserving the patient's hearing as long as possible. But allowing the tumor to grow presents its own set of problems, since if it gets too big, it is likely to impact the facial nerve, which can lead to paralysis. So the decision comes down to considering potential damage to the facial nerve versus the possibility of maintaining years' more hearing for the patient.

It was to overcome the deafness that results when the auditory nerves on both sides have to be severed that scientists at the House Ear Institute set about developing a new neural prosthetic device that could be implanted directly on the brainstem, where the auditory nerve connects to the brain. The idea was to implant the prosthesis during the tumor removal surgery. In essence, they wanted to create a cochlear implant–like neural prosthetic that would bypass both the cochlea and the auditory nerve. But until they tried implanting such a device in a human, they would have no idea whether the remaining portion of the auditory system would process the input of electrodes placed on the brainstem.

It took many years from inception to creation of a device suitable for human testing, but by 1979, the first auditory brainstem implant (ABI) was finally ready. And that was when William House advised Marilyn Davidson that she needed a second surgery on her left side that would render her totally deaf. Shortly before the scheduled procedure, she received a call from House asking that she come in to discuss an experimental program the institute was about to begin. As Davidson recalled, "They said, 'We have been working on an experimental implant for the brainstem. We don't know if it will work. It has never been done before. We don't know what you will hear, if anything, or what it will sound like.'"

Yet the proposal was music to Davidson's ear. Her sister counseled strongly against her going through with the implant, but Davidson would not hear of it. "Once they opened that door, she wouldn't let them not do it," said Robert Shannon, who heads the Department of Auditory Implants and Perception Research at the House Ear Institute. "I would say she is really the person who started the whole ABI program. Once Bill House and Bill Hitselberger [the neurosurgeon who performed the surgeries on Davidson] suggested to her that they could try this, she would not let them rest until they used her as a guinea pig to try the first one . . . She's a very forceful person and once she made up her mind that she was going to be the one to push this technology along, she was not going to hear no. I think it was really Marilyn who provided the drive to make sure that it happened."

Hitselberger agreed. "What a gutsy lady," he is quoted as saying in *House Calls*, a publication of the House Ear Institute. "If it weren't for Marilyn the ABI program would not have developed. She was the ideal candidate for the first ABI because she was so insistent that we proceed with the treatment. Consider, if you will, allowing someone to stimulate your brainstem near the delicate respiratory and nerve centers in the head. Marilyn had no idea what effect the first auditory brainstem implant would have, but she was willing to go forward with it. Marilyn was the key."

As far as Davidson was concerned, "it was the only thing to do. I just couldn't not do it. I didn't even question it," she said.

So on May 24, 1979, Marilyn Davidson, then 46 years old, became the first human being to have electrodes placed directly on the cochlear nucleus —a primitive part of the brain that is the first relay station as the auditory nerve comes from the cochlea and enters the brainstem. Placing the electrodes on the cochlear nucleus was an extremely difficult operation, since the nucleus is small, is located in the middle of the head, and is protected by the temporal bone—the one that can be felt just behind the ear. The hardest bone in the body, having the consistency of a rock, the temporal bone extends almost to the center of the head. The snailshell-shaped cochlea is a hollowed-out part of this bone. William House developed the translabyrinthine surgical technique by drilling into cadaver heads until he learned the best way to get to the brainstem.

During the translabyrinthine approach, otologists, most of whom have been trained at the House Ear Institute, do the actual drilling. Observing their progress through a surgical microscope, the surgeons use long drill bits and needle-nose-like pliers to reach into the temporal bone while looking for subtle changes in bone color and density to identify landmarks along the way. Working slowly and meticulously, they take hours to reach the brainstem. Adding to the difficulty of the surgeons' task is the fact that the facial nerve is in extremely close proximity to where they must drill. Hitting it can cause irrevocable paralysis. So their task, as Shannon put it, "is like drilling through a block of marble that contains a wet noodle you want to avoid." Once the implant site is reached, a neurosurgeon takes over to place the implant on the cochlear nucleus.

The single-channel device that was successfully implanted in Davidson was rudimentary. Though it had but one channel, there were two electrodes —one being a ground to complete the circuit—which were nothing more than two wires flared at the ends to resemble small beads. The processor that captured sound and processed it into electrical impulses that were fed to the electrodes was identical to that used with commercial cochlear implants manufactured by Cochlear Corporation of Australia. The connection from the processor to the electrodes was made through a percutaneous pedestal, which protruded slightly through Davidson's scalp. To use the system, a connector had to be plugged into the pedestal.

The day after surgery, Davidson was wheeled into an operating theater on a gurney. She had star status but didn't feel comfortable with it. She recalled seeing people packed around windows outside the operating room as William House held her hand and told her they were beginning testing. A processor was plugged into the connector in her skull, and he asked her to tell him if she heard anything. "I was scared to death," said Davidson. For what she estimated to be ten minutes but "seemed like forever," she heard nothing. Then, "I heard a sound in my head like a loud foghorn." When she told House she heard a noise, he squeezed her hand as if to say, "We did it." But problems loomed on the horizon. While recovering from the surgery, she developed meningitis and had to remain in the hospital for a month. Yet she delighted in being able to hear the sounds of the hospital, such as the nurses going about their activities, and the toilet flush.

But she also noticed a strange sensation. Whenever the implant was on, her left hamstring muscle tingled and twitched. At first, she tried to ignore it and said nothing. Then, six weeks after being discharged from the hospital, she decided to mention this strange phenomenon to someone at the institute when she went there for testing. She asked if there was any connection between the twitching and the operation of the ABI. "Absolutely," was the answer. And the response was swift. Much to Davidson's distress, the processor was taken away from her, and her world went totally silent.

Throughout development of the ABI, the House researchers were concerned about just such peripheral reactions. Their fears were born of the fact that the auditory brainstem is close to nerves associated with many vital bodily functions. The eighth cranial nerve, known as the vestibulocochlear nerve, is not far removed from the ninth and tenth cranial nerves, which control the throat, breathing, and heart functions. Though the twitching of a hamstring muscle was not a major problem, if the electrodes were in fact moving around, they could eventually cause much more serious consequences, which is why when Davidson manifested secondary effects of the electrical stimulation, her external processor was immediately removed.

House experts suspected that the problem was caused by slippage of the crude electrodes that had been placed on Davidson's eighth cranial nerve, since there was nothing for scar tissue to grow into to anchor them in place. Another possibility was that the current injected into the electrodes was simply radiating farther than anticipated or that a combination of slippage and leakage of electricity was occurring.

To help resolve the problem, the House scientists turned to electrode experts at the Huntington Medical Research Institutes, in Pasadena, long known for their electrode expertise. Over a two-year period, the Huntington scientists created a second-generation ABI electrode consisting of two small platinum bands mounted on loosely woven Dacron fabric cut into the shape of a miniature paddle. The large gaps in the weave of the fabric were intended to encourage scar tissue to grow into the fabric and thereby anchor the electrodes in place. The idea was to take advantage of the body's natural tendency to protect itself from foreign objects by building a wall of scar tissue around them. The concept has proven so successful at anchoring electrodes in the body that it has been used ever since. But electrode drift was not Davidson's problem.

As the ABI devices improved and hundreds of patients received the system without breathing or heart rhythm problems, it became apparent that the tingling and twitching Davidson initially experienced was caused by incidental stimulation of a part of the brain called the inferior cerebellar peduncle, which connects the cerebellum—associated with the control of skeletal muscle—to the brainstem. Evidence pointing to the peduncle as the culprit included the fact that a number of subsequent ABI recipients, whose electrodes were known not to have drifted, also experienced tingling and twitching sensations in various parts of their bodies. Some even reported vertigo-like sensations when their systems were turned on. Since these implantees had newer systems with as many as twenty-one individually addressable electrodes, as opposed to Davidson's two, it was found that the problem could be resolved by simply turning off the offending electrodes and using only the others. The ABI developers also learned that electrical stimulation of the auditory brainstem spreads only about 2 millimeters from the electrodes, which is well within the range of the inferior cerebellar peduncle, but not far enough to reach the ninth and tenth cranial nerves that control breathing and heartbeat.

One of the more puzzling characteristics of the side effects experienced by some ABI patients was that in almost every case, the symptoms occurred ipsilaterally, or on the same side of the body as the implant. Since the motor nerves going from the brain to the body cross from right to left, and vice versa, near the auditory nerve, if they were being stimulated by the ABI, one would expect to see symptoms on the opposite side of the body. Since that wasn't the case, the ABI sleuths deducted that it was not the motor nerves being activated by the residual current. The mystery was finally solved when they realized that electrical stimulation was sometimes inadvertently activating the inferior cerebellar peduncle in reverse. Normally this part of the brain receives signals from nerves attached to proprioceptor sensors in muscles, letting the brain know where a limb is in space. But when activated, the ABI electrodes were sometimes causing a reverse flow by sending electrical stimulation down these nerves—which normally carry signals from the muscles to the brain—thereby creating the strange sensations.

All of this was not known, however, during the two anxious years that Davidson waited in total deafness and a state of depression for the new electrodes to be developed. "When all of a sudden you can't hear, your friends disappear. They don't want to be bothered communicating with a deaf person," she said. As a result, "I was really a basket case. I was so depressed, isolated, and alone." To help her cope, Davidson obtained a hearing dog she named Spot. "I felt he was my one and only friend," she said. Hearing dogs have the same rights as guide dogs for the blind in that they can accompany their masters wherever they go. Spot ran to Davidson to let her know whenever there was a knock at the door, the telephone rang, or a smoke alarm went off. Spot actually saved her life once when "a car was coming, and he got between me and the car and pushed me over because I couldn't hear it coming," said Davidson. Spot died in 1993, but Davidson did not get another hearing dog. "I couldn't replace him, he was so special," she said. "And I just didn't want to go through that hurt again." Instead, she bought a cat and named it Tinkerbelle. During her period of silence, Davidson also took a sign language course, which she said, "opened up my deaf world. Now I have both deaf and hearing friends . . . I have said many times, if one good thing came from my hearing loss it is the wonderful people that I have met because of it."

Though Davidson waited for a new ABI in silence, she did not wait silently. She was afraid that after the problems she had, the House researchers were going to give up on the project, and she was not about to let that happen. "I had the ABI long enough to know it helped me, and anything is better than nothing." So, "I pushed them," she said. "They were trying to figure out what to do because I wouldn't let go." Her efforts eventually paid off, when in 1981, for the second time, Davidson became the first person to receive an ABI. There were some anxious moments waiting to see if she would experience the same problems she did the first time around. But her new auditory brainstem implant worked without a hitch.

Then came the long, laborious process of adapting to the implant. "It was a lot of hard work getting used to the ABI," said Davidson. The sounds that she heard were far from "normal hearing," she explained. "Instead, it's mechanical hearing. You have to get used to the sounds you hear and use them. You hear sounds like a siren, and you have to say to yourself, 'I think that's a siren,' because of the situation you are in. Is that a telephone or a doorbell? You have to listen."

And she listened carefully. Shortly after receiving her second implant, she was with her sister and heard a loud noise. She asked what it was and was told it was a dog barking. "It didn't sound like a dog, it just sounded like a loud noise," said Davidson. Yet within days, "I could tell my dog was

barking. It sounded like a dog bark." Within a year, she was able to distinguish between the sound of an airplane and a helicopter, but it took two
years for her to identify a ringing telephone. Eventually, she said, "it just
falls into place. You learn what the different sounds are." And the learning
process goes on. After wearing the ABI for twenty years, Davidson was
walking down the street one day and for the first time heard what she
thought was the singing of a bird. She looked up, and sure enough, there
was a chirping bird.

The fact that she can hear the sound of her own voice is one of the
greatest benefits Davidson feels she derives from her implant. It enables her
to maintain clear speech and to modulate the volume of her voice. And
though she has trouble making out specific words, the ABI greatly enhances
her ability to lip-read. "I communicate much better with it on than when it is
off," she said. But probably the most important part of having the device in
her head is that it gives her a psychological sense of solace, which is why she
has it turned on every waking minute. And if an equipment problem arises,
"I panic," she said.

But the benefits Davidson realizes from her implant have not come
without great physical discomfort. In all, she counted eleven times she has
been hospitalized since her surgery in 1964 to remove an acoustic neuroma
from her right inner ear. Though not all of her hospitalizations were ABI
related, as a patient pioneer Davidson did have to undergo more procedures
than a typical ABI implantee. Her second NF2-related surgery, in 1979, was
to remove growths in her left inner ear and implant the first ABI. Two years
later, the redesigned electrodes were implanted on her auditory brainstem.
The percutaneous connector was later replaced with an implanted transcutaneous radio receiver-stimulator, allowing the external processor to communicate with the implanted electrodes via radio waves that propagate
through her scalp. Several years later, the implanted receiver-stimulator had
to be replaced because of equipment problems. Subsequent surgeries were
required to remove a tumor impinging on her facial nerve and another one
that grew in her forehead. Yet despite these tribulations, Davidson maintained an upbeat attitude. Or, as Robert Shannon put it, her personality has
been magnified.

Shannon, a large man with a white beard and a Santa Claus twinkle in his eyes, explained that he considers NF2 to be "a personality multiplier." Of the many sufferers he has known over the years, he said, "some people are defeated by it. They sink into despair, and it's easy to see why. Others are sort of mad at everybody, 'Why me?' And then there are people who shine through. They have such strong and brilliant personalities." Davidson is one of them. "What makes this project so satisfying and fulfilling is being able to meet people like that. I'm not sure I would come across people like that, who are so inspiring, in my otherwise academic life."

From the time Davidson received the first ABI in 1977, to October 2000, when the FDA approved the ABI for commercial use, the device was experimentally implanted in about ninety patients, 80 percent of whom realized some benefit from the system. The implant received a big boost in 1992, when the House Ear Institute teamed with the Cochlear Corporation of Australia to produce the ABI. Cochlear added several improvements to the device, which included increasing the number of channels to twenty-two, the same number found in the cochlear implant it produces. Since FDA approval, hundreds of patients have received multichannel ABIs.

For the most part, however, ABI patients do not experience the same sound quality as cochlear implantees. One possible reason is that ABIs are implanted farther up the body's own sound-processing network and therefore cannot utilize as much of the natural hearing apparatus. Additionally, the auditory brainstem is not as neatly laid out tonotopically as the cochlea. Complicating matters further is the fact that the cochlear nucleus in the brainstem is smaller and more complex than the cochlea. Only 2 millimeters by 8 millimeters, the nerve, which is located right in the middle of the head, splits into three distinct parts in the brainstem. Each of these sections has its own frequency-processing arrangement, from high to low pitch, which is not nearly as well understood as it is in the cochlea.

To ensure that the ABI electrodes are placed where they will produce the most sound for the recipient, an electrophysiologist measures brain waves from the surface of the patient's scalp during implant surgery. To test electrode placement, small jolts of electricity are fed into the electrodes as they are moved around the cochlear nucleus. Spikes on an oscilloscope indicate that the auditory portion of the brain is responding. The stronger the response, the better the placement. Yet this is not a precise science, and on a structure where a millimeter can make all the difference, it is still something of a crapshoot as to which nerve fibers will be activated. This lack of placement precision is compounded by the fact that when the electrodes are energized, the current propagates through the nucleus to many of the thousands of nerve fibers it contains. Precisely which of them will be activated is not known until the implant is in place and turned on and the patient responds. It is then up to the technician who customizes the external processor for each patient to make the most of the functioning electrodes.

In their quest to improve on this system, researchers came upon the idea of penetrating the surface of the cochlear nucleus to more precisely target specific nerve groups. Thus, the penetrating auditory brainstem implant, or PABI, was born. It is the product of a marriage between scientists at the House Ear Institute and engineers at the Huntington Medical Research Institutes, the same union that created the ABI. The PABI utilizes needlelike microelectrodes that pierce the cochlear nucleus so that specific groups of neurons can be targeted.

The PABI was also intended to prevent the drifting of electrodes following surgery, which is thought to account for the fact that up to 10 percent of ABI recipients derive no benefit from their implants. This is primarily because the removal of a large tumor can play havoc with the anatomical landmarks surgeons use to place the electrodes. "Sometimes during the surgery the anatomy is just a mess. When they've taken out a 4-, 5-, or 6-centimeter tumor, what's left is just hamburger. It's very hard to find where anything is," said Shannon. In some other cases, surgeons find that the tumor has created an enlargement of the area in which the electrodes are to be placed. Though they try to close off the enlargement with body fat or Dacron, the electrodes can still shift before the normal scarring process locks them into place.

The PABI system includes twenty-two electrodes, eight of them the needlelike, penetrating variety, while the other fourteen are the same surface electrodes used in the ABI. By incorporating both types of electrodes into the system, the PABI developers are hedging their bets. Should the penetrating electrodes not prove successful, the patient still has the conventional ABI surface electrodes, which are known to work. The first person to receive a PABI was a 22-year-old Utah woman implanted in July 2003. Her implant was considered a qualified success, in that only one of her eight penetrating electrodes was operational. "As best as we can guess, we didn't place them in quite the right place," said Shannon. But by combining the functional penetrating electrode with the surface electrodes, the patient is able to hear some sounds, and her ability to do so is increasing with time.

The second PABI implantee, Molly Brown, experienced much greater success. Brown, who was 43 when she was implanted shortly before Thanksgiving 2003, hails from a small town in Washington State near the Canadian border. An outgoing mother of three, Brown can hear a relatively wide range of frequencies with the penetrating and surface electrode combination, ranging from a deep bass note to one she described as the highest note on a piano. This is significantly better than the sound quality realized by most ABI implantees, who tend to hear more muffled bass sounds and cannot perceive higher pitches.

When PABI recipients are sent home following surgery, their new processors are programmed with three different protocols, or maps, which control how frequency information is assigned to each electrode. One map activates just the penetrating electrodes, another just the surface electrodes, and the third stimulates combinations of both. The patients are encouraged to use each configuration to determine which works best for them in given situations. They are asked to keep a diary of their experiences with each of the maps and then return to the House Ear Institute every three months for a tune-up of their processing software based on their experiences.

For Molly Brown, who receives sound-inducing stimulation from five of the eight penetrating electrodes, the extent of her ability to hear with her new implant was immediately apparent. As soon as her processor was turned on, she was able to discern speech more clearly than the ABI recipients who had preceded her at the institute. In a standard speech-recognition test given during her early days with the implant, Brown was able to repeat back 14 percent of the words in sentences spoken to her without any visual cues. "Some of the other ABI patients have been able to do that after a year or two, but no one could do it the first day of stimulation," said Shannon. Even cochlear implantees usually take longer to recognize speech that well, although by the end of the first month or two they can average about 90 percent recognition. Though 14 percent recognition doesn't sound like much, with 50 percent recognition one can speak on the telephone. And even normal-hearing individuals tend to fill in a lot of the blanks when conversing in noisy environments, like restaurants, and don't even realize they're doing so.

MOLLY BROWN'S JOURNEY TO PROFOUND DEAFNESS began when she was 22 years old and first noticed that she was having trouble hearing telephone conversations with her right ear. A busy mother of two youngsters, she ignored the problem until the hearing in her right ear became so severely diminished that it demanded attention. Her first physician told her she had poorly formed sinuses. He gave her antibiotics and sent her home but not before he made an ominous observation. "I can still recall him holding the X-rays up and saying quietly, 'I see a shadow here,' but that was the extent of my appointment," said Brown.

Then the hearing in her left ear began to diminish as well. This time, she went to an ear, nose, and throat specialist who told her he was fearful she had an acoustic neuroma. A computed axial tomography (CAT) scan verified his suspicion, although the growth proved to be on her left auditory nerve, where her hearing was better than on the right side. Yet the tumor had to be removed, and this required two surgeries at the Virginia Mason Medical Center in Seattle. The first took place in April 1984 and had to be halted after twenty hours, when her vital signs began to falter. Brown underwent a second surgery to remove the remainder of the tumor a few months later. The procedure was successful, but she was rendered totally deaf on her left side, and the hearing in her right ear was marginal.

Early the following year, Brown became pregnant with her third child, a son born in November. One day when he was 3 weeks old, she slept until 6 in the morning and arose happily, thinking her young son had slept through the night for the first time. But her joy was short-lived. The baby had cried, but Brown did not hear him; the hearing in her right ear had disappeared overnight. Tests showed another growth, this one on her right auditory nerve. That neuroma too was removed, and while her auditory nerve was saved, her hearing wasn't. Brown was left in a world of total silence. Although neuromas on both auditory nerves are a hallmark of NF2, it wasn't until 2003, over twenty years after her first symptoms, that a diagnosis of NF2 was confirmed. One reason for the delay in diagnosis may have been that even though NF2 is typically hereditary, she had no family history of the disease.

Until the removal of the second tumor, Brown had been a gregarious person who, in her own words, "loves to talk. I am very social. Sometimes too social." Yet, plunged into silence, she went into a world of her own, becoming "almost a recluse. I was devastated and stunned at the quietness of my world. I remember asking my husband if things still made sound, or if the whole world was silent to everyone. I cried so often, I became so depressed. It was just an awful time. I was terribly lonely. I felt like I was all alone, all by myself. Even in a room packed with people and noise, I just got nothing of what was said, thus I tended to shy away from anything involving more than one or two people," she said.

Two years went by in silence until Molly's mother suggested she look into having a cochlear implant in her right ear, where the auditory nerve remained intact. She seized on the suggestion and in short order received an implant. "Oh, how happy I was the day it was turned on!" she said. "I could hear sound again. It was just wonderful." Though she experienced some residual facial stimulation from the cochlear implant, it served her well for eighteen years. Then, in 2004, she began to experience lightninglike jolts of excruciating pain in her upper right jaw that would literally knock her to her knees. "It felt exactly as though I was being shocked in a tooth, about a million volts worth. I've had a lot of different types of pain in my life, but nothing, not anything can compare to it," said Brown. She went to her dentist, who crowned a tooth, but to no avail. The pain persisted, so she went to see a neurologist, who suspected a trigeminal neuralgia, which is pressure on the nerve that supplies several parts of the face.

To find the exact location of the pressure, a magnetic resonance imaging scan had to be done. But to do a magnetic resonance imaging (MRI) scan, which utilizes high-powered magnets, the metallic receiver-stimulator embedded in her skull had to be removed, rendering the cochlear implant useless. "Here I was again without hearing," said Brown. Yet that was the least of her worries. The MRI showed that another tumor had grown on her right auditory nerve. This time, its size and location would require severing the auditory nerve during surgery, and a cochlear implant would be of no use to her even if the receiver were replaced. Molly's neurologist told her that he could remove the tumor, but if there were to be any chance of her hearing again, she should go to the House Ear Institute in Los Angeles.

She took his advice and went to House, where she met Derald Brackmann, an otologist and neurologist who specializes in diseases of the ear, facial nerve, dizziness, and acoustic neuromas. He told Brown that she was a candidate for an ABI but that she may want to consider being a test subject for the new PABI. Brown leaned toward the PABI but reserved her decision until she met with the audiologist who would be working with her to customize her system following surgery. Having had a cochlear implant for many years, she knew how important a good audiologist is. "I did not want to have to work so closely with, let's say, some jerk audiologist," she said. She hadn't told anyone of her likely decision until she met Steve Otto, the House audiologist who tunes the processors for ABI and PABI implantees. She made her decision on the spot. "I actually decided to do it right when I met him," said Brown. "He has been simply fantastic from day one." A religious woman despite, or perhaps because of, her hardships, Molly said, "I totally made up my own mind . . . I prayed and spent much time alone on it. It was just clear to me."

Despite the fact that "I full well knew of the possibility that it would not work, I really had nothing to lose—zero, nada. My hearing was completely gone. I was not afraid of electrodes on my brain and into my brain, not one bit. I was ready and willing, eager even." Brown's greatest motivation for choosing to receive a PABI was to further the research. Knowing that her disease was hereditary, her greatest fear was that her children might develop it, and she wanted to do whatever she could to further the technology should one of them need it.

So in November 2003, Molly Brown became the second human to have penetrating electrodes implanted in the auditory nerve of her brain. The surgery went well, although because of the involvement of the trigeminal nerve, she came out of surgery with an inability to move many of the muscles on the right side of her face or to close her right eye. As a result of the facial paralysis, the right side of her face looked like it was sagging, and made eating and talking difficult. Brackmann, who performed her surgery, told Brown that between six and twelve months after surgery she would begin to see an improvement in her paralysis. "Then, sure enough, one night around eight months post-op, I was looking closely in the mirror and there was ever so slight a movement," said Brown. "I actually could not sleep at all. It has continued to improve nearly every day. I had a dimple on that side, and it is even showing up on occasion."

The inability to close her right eye compounded her postoperative problems. During another procedure, performed a week after her PABI implant surgery, a spring was placed in her eyelid, returning her ability to blink. With poor vision in her left eye caused by a childhood fireworks accident, Brown was terrified of going blind as well as deaf. Even after the spring was implanted, she could not produce tears, so despite drops, her eyes were constantly irritated. She then developed a painful ulcer on her cornea from the spring rubbing against it. Subsequent reconstructive surgery has enabled her to blink on her own again, improving both her appearance and vision.

One of the hardest things Brown has had to endure are the "stares and just really stupid comments" she has received as a result of her facial paralysis. "I do not care at all what people look like—it really is what's inside that counts," she said, acknowledging that "I was never a Miss America, but the stares and comments wear on me."

In spite of hardships that would break a lesser person, the real payoff for Brown came in January 2004, when she returned to House to have her audiologist Otto, in her words, "turn me on." Brown knew exactly what to expect. Even so, she was surprised at how calm she was, considering no one knew if her PABI would work. Once she was hooked up to a processor and Otto did, in fact, turn her on, there was a period of thirty seconds that seemed like forever during which she heard nothing. Then, she said, "I clearly heard the 'beep-beep' sound programming makes. It is nearly an identical sound to my cochlear implant programming." With that sound, her inner composure disintegrated. All she said was, "There." But despite her calm demeanor, "inside I was just in a frenzy! God heard ten thousand thank-yous in twenty seconds!"

Otto then went about the tedious procedure of testing and tuning each of Brown's electrodes for optimum performance. This process requires the audiologist to assign the various pitch regions in speech to the appropriate electrodes based on Brown's response to test inputs. In this trial-and-error process, Otto stimulates the electrodes one at a time and measures the wearer's threshold for volume by first raising the current to the electrode until it produces a sound that can barely be heard. He then raises the stimulation to a loud volume and measures the dynamic range between the two points. Once all of the electrodes are measured for volume, he goes back and does the same for pitch. He then removes that electrode from the mix and asks the patient which of the remaining electrodes gives the highest pitch, and so on until the one that gives the lowest pitch is identified. He then sets the processor to separate environmental sounds, including speech, into varying levels of pitch to coincide with the pitch levels associated with each electrode, and he sends those signals to the respective electrodes.

After three days of such testing and tuning with Brown, Otto felt she and her processor were ready. "He hit the button and said things like, 'So how does that sound? Testing one, two, three. What does my voice sound like?' " recalled Brown, who was immediately able to make out much of what he said. "I was about ready to jump into his arms and kiss him! Though I'm sure he is glad I didn't," she said. "I felt his voice was the most beautiful voice in the world. The sounds were so similar to my cochlear implant I just could not believe it."

Enthusiasm aside, Brown acknowledged that what she hears "is mechanical hearing, and I highly doubt that you would be all that pleased with it. But for me, I was just thrilled, elated, wild with happiness . . . I have only had these *big* emotions three times in my life, and I have three kids. It is a feeling of quiet-loud contentment—just beautiful."

Brown's processor gives her a variety of control settings, including microphone sensitivity and volume, which she experimented with to determine what works best in different environments. "I am working on adjusting the sensitivity up, then maybe lowering the volume just a bit, and seeing how that goes," she said. By doing so, she can make what she hears "more harsh sounding or less, or more clear, or more fuzzy." In the process, she appreciates the remarkable nature of what is going on in her brain. "It is fascinating, and quite a treat to be able to do something so amazing as change sounds with the push of a button. I do feel bionic some days," she said.

Even well after her PABI was activated, Brown found that she had trouble sleeping, "because I am so happy and excited about it." Long after receiving her implant, Brown continues to hear and identify new things. To her, it is a challenge that she thoroughly enjoys. "I catch myself saying all the time things such as: 'That was a jet, wasn't it?' Or, 'Was that someone talking over the intercom?' They have their own distinct sounds, much as they do to you. They just don't sound at all like you would know them. But once I get it, then it seems programmed into my brain."

Two months after receiving her PABI, Brown returned to work cleaning houses, which she had done for several years prior to being implanted. "It's a wonderful job for me," she said. "It forces me to interact and try to hear. I love the people I work for, the pay is not too bad, plus I get some exercise while being paid!" The money she makes helps defray the tremendous financial burden brought on by her extensive medical care. Though the PABI implantation and all the related expenses were paid for by research grants, much of her care prior to becoming a PABI research subject was not covered by insurance. "The financial burdens are huge. You just have to look at my chronology and full well know that I am much more than a million-dollar girl. It is just astounding when I think about how much money has been expended on my head and my life!"

Insurance has come through for "most things," she said, but only after badgering by her mother. "My mom is just a bulldog on them. I swear she could get them to pay for you to have a baby," she told a male acquaintance. Even so, Brown said the bills she receives "would scare the pants off you. I am just so nonchalant about getting a bill for $100,000-plus anymore." To cover them, she and her husband have taken out loans and she has received assistance from her parents and brothers, though she has never asked for it.

"I am very fortunate, indeed," said Brown. "Having such endless encouragement and support is another reason I do so well. I am thankful beyond words." And as for the implant itself, "I am so very happy with it. It is not at all perfect, but it is *something*. I am not nearly as lonely as I was with no sound. If you could say there is something such as a medical miracle, this would have to come very close. I'm just so thankful."

A personality multiplier, indeed!

⟞⟊⟍⟊⟞

YET THERE HAVE BEEN SETBACKS. During follow-up testing, Brown has never duplicated the word-recognition success rate she had during her first test. And, though her initial abilities with the implant were exceptional, over time her rate of improvement has slowed to the point that her ability to hear with the PABI is no better than that realized by high-end ABI users. "That's been a disappointment," acknowledged Shannon. The PABI developers have suffered other setbacks as well. Of the three individuals who received PABIs after Brown, only one can make use of the penetrating electrodes, even though they all have functioning surface electrodes. In one of those implantees, X-rays have shown that the penetrating electrodes popped out of the cochlear nucleus. And, as if that wasn't enough, Shannon and his team came face to face with the frequently circuitous, strained, and frustrating path of medical advancement, when late in 2004 Shannon took a trip to Verona, Italy, and made an astounding discovery.

He went there because he had heard from an Italian physician named Vittorio Colletti, chairman of the otolaryngology department at the University of Verona Medical Center, who claimed that he was implanting ABIs in individuals whose auditory nerves were damaged by causes other than NF2, such as accidents or other diseases, and was achieving much better results than those realized by the House ABI patients. In fact, said Colletti, he had implanted about forty individuals, and most of them were hearing as well as people with cochlear implants. The auditory implant community was skeptical. So at Colletti's request, Shannon went to Verona to test Colletti's patients himself. What he found "just blew my socks off," said Shannon. "I couldn't believe that with the same device we've been using for all those years in 180 patients, more than half of his patients are performing better than any of ours . . . He has people talking on the phone. It's unbelievable. We've never seen anything even remotely close to that performance. If he had an outcome like ours in any of his patients he would consider that disappointing."

In one fell swoop, Colletti had knocked roughly twenty years of PABI research and development on its ear. The House and Huntington Medical Research Institutes scientists developed the PABI based on the assumption that the marginal results realized with the surface electrodes of the ABI were caused by their inability to precisely target specific frequency-processing neurons. They hoped that the penetrating electrodes of the PABI would rectify that shortcoming. By obtaining better ABI results than Shannon and his colleagues were able to achieve with either the ABI or PABI, Colletti threw that entire theory into question. Instead, it is believed that Colletti realized superior results specifically because he implanted ABIs in patients who do not suffer from NF2, and who therefore did not have tumors on their auditory brainstems prior to implantation. The theory holds that it is damage done by the tumors themselves that prevents the ABIs from performing to their full potential. This conclusion has gained even more credence from the fact that Colletti also implanted ABIs in NF2 patients and realized results no better than those of the House ABI implantees.

The House brainstem implant scientists have taken a many-pronged response to Colletti's astonishing results. For one thing, their thinking on the risk/benefit ratio of ABI implantation has changed. Initially, they felt that because implanting ABIs and PABIs is a risky procedure and the benefits had proven to be marginal, such implants should only be done on people requiring surgery on the brainstem for other reasons, such as the removal of NF2 tumors. With Colletti's results, that equation has changed. House researchers soon began themselves to seek candidates for ABIs who had nonfunctioning auditory nerves that were damaged by causes other than NF2 tumors. They also decided to try implanting NF2 patients whose tumors have not yet impinged on the auditory nerve. Prior to the Colletti findings, they waited until tumors had already significantly impacted the patients' hearing to do the surgery.

Meanwhile, they are not ready to abandon the PABI approach. In 2005, the House researchers received FDA approval to implant ten more patients with the experimental device. For this round of testing, they have made slight changes in the electrode design. They will also take greater pains to stabilize the penetrating electrodes during surgical insertion and closing to prevent subsequent movement.

But even if the PABI concept proves less than fully effective for people whose auditory nerves have been damaged by tumors, all of the work that has gone into its development will not be for naught, since the electrode technology used for these implants has found its way into other neural implants. The possible failure of the PABI to improve the performance of the ABI would, of course, be bad news for NF2 patients, but it would not be a hopeless situation. "If the cause of the poor performance is NF2, and penetrating electrodes don't seem to be overcoming the problem, we are going to have to take a different tack," Shannon said. "That is not what we want to find out, but that's why you do research. You have to figure these things out as you go."

The next attempt to improve ABI results for NF2 patients will be to implant electrodes on the auditory system's next relay station in the brain, the inferior colliculus, which is several centimeters farther up the brainstem from the cochlear nucleus—the equivalent of yards in the minute scale by which distances in the brain are measured. The inferior colliculus is not damaged by NF2 tumors and, about the size of an olive, is larger and easier to implant than the cochlear nucleus. Additionally, its tonotopic layout is better understood. So why was it not implanted from the outset? "Because you're already in there [at the auditory brainstem] for tumor removal, and philosophically we felt it is better to implant and stimulate the most peripheral location," explained Shannon. Looking to the future, Thomas Lenarz and Minu Lenarz, a husband and wife team at the Medical University of Hanover, Germany, have an inferior colliculus auditory implant project under way, while another group at the University of Michigan, in Ann Arbor, is also investigating such implants.

Another important part of the House research program is aimed at gaining a greater understanding of how the brain takes external stimuli, from one's own sensory organs or from a neural prosthetic, and puts them all together into coherent information. Such knowledge not only expands humankind's basic understanding of the operation of the brain but also enables those creating neural prostheses to develop systems that provide stimulation that is readily processible by the brain.

A greater understanding of how the brain processes sensory information may eventually lead to the prevention of disabilities, such as deafness, from occurring in the first place. In the meantime, the expertise gained through the creation of auditory neural prosthetic devices—which in the form of cochlear implants, were the first sensory implants to find widespread clinical acceptance—has gone a long way toward aiding in the development of other neural prosthetic devices, such as retinal implants designed to enable the blind to see. Following the developmental curve set by the auditory prostheses, such implants are now being implanted experimentally in numerous people, with the potential to become relatively commonplace as time and science march on.

10

In the Eye
of the Beholder

In a small, windowless room chockablock with ophthalmic equipment at the Doheny Eye Institute on the University of Southern California's Medical Center Campus in Los Angeles, a slight, 76-year-old woman sits in a chair facing a blank wall. Although the room is dark, she is wearing a pair of sunglasses as well as a patch over her left eye.

Connie Schoeman is totally blind, yet she is able to tell biomedical engineer Arup Roy whenever she sees a square-shaped light image appear on the wall in front of her. Operating the computer that controls the light display, Roy flashes the image a total of ten times, and Schoeman correctly identifies all but one. She can do so thanks to a minuscule spatula-shaped sliver of plastic containing sixteen round platinum electrodes attached to her retina, the tissue paper–thin membrane that lines the inner wall of the eye.

The images Schoeman sees are transmitted to the retinal implant in her eye by a tiny camera mounted on the bridge of her glasses. A wire leads from the camera to a small circular object that sits firmly on her scalp just behind her right ear. It contains a miniature radio transmitter that is held in place by a magnet attracted to a receiver-stimulator implanted in a hollowed-out portion of her skull. Under the skin and musculature of her scalp, a small cable made up of sixteen tiny wires runs from the receiver-stimulator to her right eye, where it is attached to the four-by-four electrode array on her retina.

Schoeman now readies herself for a more difficult test of her neural prosthetic device. As Roy begins this test, a tone chimes periodically. Sometimes the tone is accompanied by the projection of light onto one of four corners of an imaginary square on the wall. Other times, no light accompanies the tone. Schoeman is asked to determine whether the light has come on, and if so to point to its location. With each tone, her head sweeps in exaggerated up-and-down and side-to-side motions as she scans the wall looking for the light. This motion is made necessary by the fact that the electrodes inside her eye offer no peripheral vision. They are actuated only when the camera on her glasses is aimed precisely at the light. A slight movement one way or the other and they go off, requiring Connie to immediately turn her head back toward the light. It can be a frustrating exercise, much like an amateur astronomer trying to find a planet or star in the night sky with an inexpensive telescope. The light-in-the-corner test is run in two sets of ten tones. During the first set, Schoeman correctly identifies and locates the light nine out of ten times. During the second, her accuracy falls to six out of ten, perhaps because of fatigue.

In previous tests, Schoeman has been able to identify a paper plate, a cup, and a knife, not because they look like what they are to her but by the number and configuration of the points of light she sees when the electrodes in her eye are activated. She was told ahead of time that various white eating utensils were placed in front of her on a black background. She was then asked to identify what they were.

"I know it's a paper plate because there are more dots of light for the plate than there are for the cup, and the knife has dots going down like in a column," she said. In the case of the plate, she doesn't see a circle because she does not have enough electrodes in her eye, but she does see dots on either side, much like what a dot matrix printer with only sixteen dots would produce. Even though Schoeman can see only light dots that resemble incandescent bulbs, that is a lot more than the nothingness she saw prior to being implanted with a retinal prosthesis.

Schoeman was blinded by retinitis pigmentosa, an inherited degenerative disease that affects the photoreceptor cells of the retina—the rods and cones. The retina, which processes light, lines the back wall of the eye and is actually an outgrowth of the brain. The rods and cones, named for their shape, react to light and begin the process of converting light to electrical signals, or action potentials, which is completed as the signals generated by the rods and cones pass through two other layers of cells in the retina. The human eye has about 100 million rods and 3 million cones.

As light enters the eye, it first passes through the cornea and then the iris, which controls the amount of light that enters, much as the diaphragm of a camera controls the aperture opening. The lens of the eye then focuses the image and projects it upside down onto the retina, just as a camera's lens reverses the image it projects onto film. When processing the image, the brain flips it right side up. Having a damaged retina is like having bad film in a camera. The image that passes through the lens will not register.

Strangely enough, the rods and cones, which begin the process of converting the light to electrical signals, are located at the back of the retina, which being transparent, allows light to pass through. After the photoreceptors convert the light to analog electrical signals, the signals are passed back up through the retina to the next layer, the bipolar cells, which process them further and forward the impulses to the amacrine cells. The impulses are fully digitized finally by the ganglion cells that feed the optic nerve—a bundle of about 1.2 million nerve fibers that can fire at up to two hundred pulses per second. The optic nerve exits the back of the eye, just above the center of the retina, and runs to the lateral geniculate nucleus in the brain, where the optic nerve fibers make synaptic connections with optic radiation fibers. These fibers carry the signals to the visual cortex at the back of the brain, where they are processed into recognizable images.

Retinitis pigmentosa is a disease that slowly degenerates the rods and cones, first robbing its victims of their peripheral vision, then slowly closing in and shutting down their central vision as well. The success of the retinal prosthesis is based on the belief that the other layers of cells in the retina remain intact even though the rods and cones are destroyed. The extent to which this is so remains a point of dispute among experts. If enough of the other parts of the retina remain intact, the implanted electrodes may be able to replace the damaged rods and cones by converting light to electrical impulses and starting them on their way through the rest of the retina. Of course, the sixteen electrodes in Schoeman's implant are a poor substitute for millions of photoreceptors in the healthy retina, and no one expects that they will provide much in the way of functional vision. But it is a start.

One of the things researchers hope to do through tests conducted on Schoeman and others like her is answer the retinal implant million-dollar question, namely, how many electrodes will be required to produce useful images? Their goal is to increase the number of electrodes into the hundreds and someday even the thousands, but will that be sufficient to create useful vision? Developers of retinal implants take heart from the experience with cochlear implants that have given hearing to many thousands of profoundly deaf people. When cochlear implants were being developed, some thought at least a thousand electrodes would be needed to create coherent sound, yet six electrodes proved to be of great help to some patients. This points to the incredible plasticity of the human brain, which allows it to take a somewhat crude sensory input generated by a manmade device and make sense of it. Yet how many electrodes it will take to provide useful vision remains anybody's guess.

This speaks to another crucial question retinal implant developers must grapple with: How much sight is sufficient to make the effort and expense worthwhile? Is being able to navigate on one's own sufficient, or should the ability to recognize faces and read be the criteria? Some say such a system must provide more than a cane, a guide dog, and Braille before it will be of any real use to people who are blind. This, however, begs the question of psychological benefit that the few test implantees seem to derive from being able to perceive any light. As Schoeman put it, "With something like retinitis pigmentosa, where there has never been anything that could be done to help people, this is going to offer an opportunity, maybe not to get complete vision back, but something. And something, I tell you, sure beats nothing."

Clearly, there are many complex questions that must be tackled before retinal implants become widely available, some of which will be answered by the testing of people like Connie Schoeman, who received her retinal prosthesis on March 12, 2003, when she was 75 years old. A New Englander from Providence, Rhode Island, and Norwich, Connecticut, she attended Pembroke College—then the women's branch of Brown University in Providence —for three years before transferring to the University of Southern California, where she received a degree in clinical medical technology. She went to work as a laboratory technician, a job that required the daily use of a microscope. Several years passed before she had any inkling that there was a problem with her vision. In an odd twist of fate, her first job was at the county hospital directly across the street from the Doheny Eye Institute, where she now undergoes testing of her retinal implant.

The first sign of trouble with Schoeman's vision came while she was driving home from work one evening when she was 27. She drove into a parked utility trailer that she felt she should have seen. Thinking that something was amiss, she went to an optometrist, who told her she had night blindness and should not drive at night. Two years later, she was involved in another auto accident that convinced her something was seriously wrong. A woman backed out of a driveway into Schoeman's passing car. While the accident was judged to be the other driver's fault, Schoeman knew she should have seen the car coming out of her peripheral vision.

This time, she went to see an ophthalmologist, who immediately suspected retinitis pigmentosa. When he asked if anyone else in her family had eye problems, she told him that her father had glaucoma and cataracts. Schoeman's doctor checked with her father's physician and learned that he also suffered from retinitis pigmentosa, something Connie was unaware of. Now she had the blinding disease as well. Her physician told her never to drive again. "But I drove here," she protested. "But you shouldn't drive home," he told her. Just to be safe, her husband tore up her driver's license. The diagnosis was rendered in 1956, when she was 29. She was still able to read and get around on her own, but her vision continued to deteriorate for another thirty years until she became totally blind. "There was no day that I can remember becoming totally blind," said Schoeman. "It was a very slow process."

As her sight deteriorated, Schoeman was determined not to be defeated by her impending blindness. She began preparing for the inevitable by studying Braille at the Braille Institute of America. Since she could no longer work as a laboratory technician, she also began studying for a master's degree in vocational rehabilitation at California State University, in Los Angeles. Since she was already unable to read the required texts, she employed people to read them to her, and she took notes using a Braille typewriter. When test time came, she had someone read the questions to her and she dictated the answers. She graduated in 1967 and got a job with the California Department of Rehabilitation. A driver was assigned to her so she could go to the homes of blind people to teach them Braille and safety techniques in the kitchen. She then became a vocational rehabilitation counselor helping blind people select occupations. She remained with the Department of Rehabilitation for twenty-six years.

Schoeman learned of the retinal implant tests when she attended a couple of seminars, one at Doheny, the other at the Braille Institute. She submitted her name at the first seminar and heard nothing. Anxious to be a part of the study, and typically persistent, she put in a follow-up call to the office of Mark Humayun, one of the implant's inventors, and was able to procure an appointment with him. Though she knew that nothing was being promised, she "thought there might be a possibility that maybe this would improve my vision."

In fact, each retinal implant candidate was informed at the outset of the very faint possibility that the implant would improve their vision. "They wanted to be completely honest. They didn't want to promise something and then have me disappointed," said Schoeman. "But it is so interesting, and there was a faint possibility, and hope springs eternal." That hope extends to future generations. Schoeman happily said that the hereditary disease that blinded her has not affected her children or grandchildren. "But that doesn't mean that the generation after that isn't going to get it. So if there's a way for anybody who gets retinitis pigmentosa, especially if they are my relatives, to benefit from something I did, that makes me feel good."

Within a week of surgery, she was at the Doheny Eye Institute ready to begin a rigorous testing regimen. In the years since, her ability to see more with the implant has not improved, but, she said, she is learning to see better, which seems to verify the notion that the brain adapts to make the greatest possible use of available input. "When I'm looking for the line of light, I can see it better. The more I do it, the faster I can find it," she said.

———∘∘∘———

THUS FAR, THE FEW PATIENTS WHO have received experimental retinal implants suffer from retinitis pigmentosa, though these same prosthetics would be applicable for people with other retinal diseases as well, like some types of macular degeneration. Approximately one in every 4,000 people are born with retinitis pigmentosa, while some 30 million people are afflicted with age-related macular degeneration in the Western world alone, and 100,000 new cases develop in the United States each year, making it the leading cause of blindness among people over 55. There is, therefore, a large potential market for a safe, practical retinal implant. This has led to the creation of at least two companies to pursue the development and eventual commercialization of retinal implants.

Second Sight Medical Products Inc., of Sylmar, California, was set up to advance the system that Schoeman and several other test subjects have in their eyes and to manufacture and sell eye implants if they are proven effective and approved for general use by the FDA. When that will happen is an open question, although Robert Greenberg, Second Sight's president, hopes to have a commercial implant on the market by the end of the decade.

The chip that is attached to Schoeman's right retina had its genesis in the mid-1980s, about twenty years before she was implanted, when Mark Humayun was a student at Duke University Medical School. Humayun grew up in Potomac, Maryland, near Washington, D.C., literally surrounded by physicians. His mother and four uncles are physicians, as was his paternal grandfather, though none specialized in ophthalmology, and ophthalmology was not in his sights when he decided to become a doctor. But he did carry the memory of his maternal grandmother, to whom he was close and who went blind as a result of diabetic retinopathy, a disease of the retina. The fact that nothing could be done to help her made Humayun painfully aware of medicine's limitations in dealing with diseases of the retina. Yet as he progressed through medical school, he leaned toward becoming a neurosurgeon.

Though he attended medical school at Duke, Humayun spent his third year as a medical student doing research in neurosurgery at the Johns Hopkins School of Medicine. While there, he had the opportunity to observe a surgery, performed on a seizure patient under local anesthesia, in which various parts of the brain were electrically stimulated to find the focus of the seizures. When the visual cortex of the brain was stimulated, the patient reported seeing flashes of light. This was a turning point in Humayun's career. When he returned to Duke, he did a clinical elective rotation in ophthalmology, during which he became acutely aware of the large number of people who go blind from retinal disease.

"It made me think that a lot needed to be done in this area," said Humayun. "It rekindled that flame that started burning, but was on the back burner, due to my grandmother's condition. I pieced it together and realized that I really wanted to do ophthalmology." But he felt he had to take one other step before his decision was final. He wanted to make sure that he liked the research part of ophthalmology as well, because though he did want to practice clinical medicine, he also hoped to spend some of his time in the laboratory. This quest led Humayun to the laboratory of Eugene de Juan, a young professor of ophthalmology at Duke.

The year was 1987, and de Juan was investigating ways of suppressing the abnormal growth of blood vessels in the eye caused by a number of conditions, such as macular degeneration and diabetes. The abnormal proliferation of blood vessels can lead to bleeding, which can cause serious visual problems. Much of de Juan's work involved pharmacology and cell biology, but Humayun was looking for something that involved a more hands-on approach. Vividly remembering the electrical stimulation of the brain he witnessed at Johns Hopkins, he asked de Juan if he thought there was any way the eye itself could be electrically stimulated. As luck would have it, de Juan had recently developed a metallic tack that looks like a minuscule thumbtack and is used to secure detached retinas to the wall of the eye. He told Humayun that since these tacks are made of metal and go inside the eye, placing metal electrodes inside the eye was not out of the question. Encouraged, Humayun decided to pursue the idea of returning sight to the blind through electrical stimulation of the retina. De Juan also found the idea intriguing.

One stumbling block that quickly became clear to them was that they lacked the electrical engineering expertise needed to proceed with such a monumental project. They found it in Howard Phillips, a neighbor of de Juan's who had a doctorate in electrical engineering and worked at Semiconductor Research Corporation at Research Triangle Park, in North Carolina. During many after-work back porch conversations, Phillips gave the two ophthalmologists a crash course in microelectronics. With their ophthalmic expertise and the information gained during their informal tutorials, the two physicians set about the work of creating a retinal implant. But it didn't take long for them to realize that what they learned about electronics over the back fence was far from sufficient. They concluded that one of them should return to school to gain the necessary knowledge.

Being an inveterate student, Humayun volunteered. But he wasn't just going to take a few courses, he was going to get a doctoral degree in biomedical engineering. To do so, he signed on for a program at the University of North Carolina, Chapel Hill, which gave him the flexibility to do his medical internship and residency while studying to become a biomedical engineer. As for how he managed the famously long hours residents must clock at hospitals while studying for a Ph.D., Humayun said, "It was a hardship, no doubt about it." He managed to do so primarily by studying while others were catching an hour or two of sleep. "I realized that during internship and residency you are not getting sleep anyway, so instead of getting an hour or two and then being awakened—I don't do well with these 'power naps'—I just stayed up," he said.

And as if that weren't enough, he and de Juan continued to work on the development of a retinal prosthesis at the same time. Though an implant was nowhere near ready, the two physicians began human testing in 1992. The tests were designed to answer a number of critical questions, which would determine whether their idea was worth pursuing any further. Key among them was whether the cells in the blinded retina, other than the photoreceptors, remained viable. If the entire retina degenerated along with the photoreceptors, then their work was an exercise in futility, since there would be no remaining biological circuitry into which the signals from the implanted electrodes could be fed. As a first step in answering this question, the ophthalmologists studied retinas taken from deceased retinitis pigmentosa patients. They found that though less than 1 percent of the rods and cones were intact, about 80 percent of the remaining retinal cells were morphologically undamaged. Even so, "we still had no idea what the effect of fifty or sixty years of degeneration would be on the response of these cells," said Humayun. To determine if they remained functional would require inserting an electrode probe into the eye of a living patient.

The physicians were also concerned about what the effect of the passage of current through the vitreous gel that fills the inside of the eye would be, since it is about 99 percent water. Because current spreads throughout water, their fear was that a patient might see nothing but a huge, uncontrollable flash of light. And there was also the question of what an artificially created electrical pulse would look like. Would it look like a dot, or be blue or green? Would it be something that is appealing, or would the stimulus be so noxious that the implantee would rather be blind?

To answer these questions required a live volunteer. One was found in the person of Harold Churchey, a 71-year-old from Sharpsburg, Maryland, who was blinded by retinitis pigmentosa. From birth, Churchey didn't have much vision in his right eye and had what he calls "gun barrel vision" in his left eye. Until he reached his midthirties, what little sight he had remained stable. Then it too began to deteriorate. The caliber of the gun barrel slowly decreased until he got to the point that on good days he could merely tell the difference between daylight and dark. Churchey's ophthalmologist at Johns Hopkins University's Wilmer Eye Institute knew that de Juan and Humayun were looking for someone who had been blinded by retinitis pigmentosa for their tests, so he asked Churchey if he would like to participate. It was explained to him that the tests had no chance of helping him see again, and yet he did not hesitate to volunteer. Explaining why, Churchey told of encountering a 5-year-old boy who also had retinitis pigmentosa, at his ophthalmologist's office. Knowing what the future held for him, Churchey's heart went out to the boy. "Even though I might be over the hill, if I can help some young person, I'm for it, so long as the good Lord gives me strength," said Churchey.

What he was "for" was having the two Duke ophthalmologists sink a single platinum electrode through the white of his right eye and slowly move it downward toward his retina as they injected small jolts of electricity. The first test took place at Duke on September 17, 1992. As de Juan and Humayun manipulated the electrode in Churchey's eye, they anxiously waited to hear him report what he could see, if anything. The entire future of their project hung in the balance. For what seemed like an eternity, he reported nothing.

"There was pin-drop silence because he didn't see anything for twenty minutes or so, and we were frantically checking every circuit, every wire," recalled Humayun. "It's almost like when the space shuttle reenters the atmosphere and there is this period when there is no communication with the crew. You don't know if everything is OK. Twenty minutes in the operating room felt like eons, and then, he mentions, 'Oh, are you talking about that small little light blinking on and off in the distance?' We were like, 'Yes!' " To turn the blinking light into a constant light the researchers cranked up the stimulus pulse so that it flickered fast enough to look continuous to Churchey. This mimicked the healthy visual system in which the cells fire then briefly rest. While the cells are resting, the brain retains the image it has just received until the next image, in the form of electrical stimulus, arrives so that the perception is of a continuous image. It is this image retention that allows the flickering of individual pictures on film passing through a projector to look like unbroken images on a movie screen.

Describing what he saw that momentous day, Churchey said, "It was just a little light. That's all it was." But, he added, "from the time they put the probe in my eye, I knew they were on the right track." To make sure the light was not a figment of Churchey's imagination, the ophthalmologists turned the stimulating electricity off and on, and each time they did so, he was able to see the light go out and come back on. He also saw the small spot of light move as they moved the probe around in his eye. For approximately one hour, he watched and reported what he saw as the physicians altered the current and moved the probe. They then removed the probe from his eye, and he was wheeled from the operating room. That hour was enough to stoke the fire in the two physicians that has fueled their ongoing research and development for many years since.

"That was the most defining point in my life," said Humayun. "There was a sense of relief and gratification. On the other hand, as Mr. Churchey was being wheeled out of the room, I realized that we could not go back now, that we had to build a device because now it was clear that the patient could see, that this idea works. So, we felt this relief, and we also felt the incredible burden that we had to take this road. If he hadn't seen, well, it would have been a great experiment, and I would have been crushed to some extent, but at least the book would have been closed. But instead it opened up a whole new chapter. We knew we had to build a device."

Shortly after the first Churchey test, Eugene de Juan moved from Duke to the Wilmer Eye Institute at Johns Hopkins, where he became a faculty member. So the two could continue their retinal work, Humayun followed not long thereafter. Their second human test was on another patient in whose eye they inserted a probe, this time containing three electrodes with an edge-to-edge separation of 300 microns—the width of a few hairs—to determine if the patient could see individual spots of light. That too was successful. They conducted a similar second test on Churchey in 1994, and then two years later they again slid a probe into Churchey's right eye, this one with twenty-five electrodes on its spatula-shaped tip. They attempted to project the letter U onto Churchey's field of vision, but they picked the wrong letter. Since they were generating individual spots of light, it took on a pixelated look, and without the rounded edges, it appeared to be an H to Churchey.

De Juan and Humayun tested various probe configurations on a total of seventeen patients, all of whom knew full well that putting a probe into their eyes had absolutely no immediate value to them. "All of them are very much part of our team and are truly altruistic. After all, it's eye surgery and you're not getting any benefit out of it," said Humayun. But the benefit of their contributions to the development of a retinal prosthesis was immeasurable. These experiments proved that electrical stimulation applied to the retina could cause people who were blind to see light spots, indicating that the cells other than the photoreceptors in the retina were still functional and that the light generated was spatially discernable and palpable. They also garnered invaluable information about how strong an electrical charge is needed to create light spots. Yet the information gathered during these early human tests was only the first step in responding to a long list of questions that needed to be answered and barriers that had to be overcome before a retinal chip could become a reality.

A major challenge was to build a device that the eye can tolerate for many years and that will not be damaged by the saltwater environment inside the human body. As one expert put it, implanting a neural prosthetic device inside the human body is "like taking a television set and throwing it into the ocean and hoping it still works." Heat dissipation was another big question. The system had to be designed so that the amount of power required by the implant would not create excessive heat for fear that it could cook the eye, much like an egg in boiling water.

—⊸⊷⊶—

BECAUSE HAROLD CHURCHEY WAS THE first person to submit to testing by Humayun and de Juan and because he willingly submitted to many additional tests, the two visual prosthetic pioneers promised that he would be the first person to receive a permanent retinal implant when it was ready for human testing. By 2002, when that was the case, the ophthalmologists had changed their research venue to the University of Southern California Medical Center's Doheny Eye Institute. But that didn't prevent them from making good on their word, though Churchey lived in Maryland. When they asked him if he still wanted to be the first person to receive the permanent retinal chip, his response was immediate and affirmative. It wasn't long before he and his wife Eva were on a plane to Los Angeles, and within a matter of days, on February 19, 2002, he underwent seven hours of surgery involving three teams of surgeons to implant the artificial retina in his right eye.

According to Humayun, who led the ophthalmic team doing the actual placement of the chip on the retina, one of the reasons for the lengthy surgery was that the receiver-stimulator that processes input from the camera and sends signals to the electrodes in the eye, is a modified cochlear implant and must be implanted in the skull behind the ear. Eventually, engineers hope to reduce that receiver-stimulator unit, now about the size of a woman's compact case, down to something that can be placed inside the eye along with the chip containing the electrodes. Humayun estimated that once that is accomplished, the entire surgical procedure will take no more than two hours. That may not be very far into the future, since such small devices were already undergoing animal testing as he performed the first implant on Churchey. In the meantime, seven hours is the norm for the surgery, which is performed under general anesthesia.

The first step is for an ear, nose, and throat surgeon to place the receiver-stimulator behind the ear, just as a cochlear implant is positioned. A thin cable containing wire from the processing unit is run underneath the skin and muscle toward the eye. A team of ocular plastic surgeons then runs the cable behind the eye, after which Humayun and his ophthalmic team take over. First they remove the sclera, or white of the eye, which is akin to skin, and then route the cable around to the back of the eyeball. The wire is sutured to the exterior wall of the eye at each quadrant of the eyeball, allowing it to move with the eye so that strain is not placed on the electrode array when the eye moves. The final step before closing is to place the electrode array through the eye wall and attach it to the retina using metal tacks the size of two human hairs. The white of the eye is then sutured back into place and the other incisions closed up.

Churchey remained in the hospital for only one night following surgery and was allowed to recuperate for about two weeks before he was ready to test his new implant. The first test was run in a room at the institute packed with about thirty people anxious to see this major milestone, the culmination of roughly fifteen years of research and development on the part of de Juan and Humayun, and by now many others as well. An external transmitter was placed over the receiver embedded in Churchey's skull behind his right ear, and he was handed the camera that was to feed visual information to the sixteen electrodes sitting on his retina, as it was not yet mounted on a pair of glasses. A large letter L was projected on a wall, and Churchey's system was turned on. The tension in the room was reminiscent of the first time de Juan and Humayun sank an electrode through the white of Churchey's right eye ten years earlier.

"We asked him to point the camera toward the wall and tell us what he saw, and he starts shaking his head," said Robert Greenberg. "Everybody in the room was thinking, 'This is terrible, he is not seeing anything.' You could see the whole room was really deflated when he started shaking his head. And then he says, 'All I see is a line that goes up and down and one that goes across. He traced the L out with his hands. You could see the whole room basically burst. Everybody was ecstatic, obviously. We asked him what it is, and he said, 'I don't know, it just looks like an L.' "

Churchey remained in California for the next nine weeks, during which a wide range of tests were conducted so the implant team could learn about the system in his eye and Churchey's brain could adapt to the unnatural input and learn to better recognize images produced by his implant.

By far the most dramatic of the early tests took place when Churchey was first given a portable processor and asked to go out into a sunlit day to see what he might see. As traffic went by, "I didn't know if it was a Model T or a Sherman tank, but I knew there were vehicles going past," said Churchey. For the first time in many years, he could "see" the vehicles and even determine the direction of their travel by observing the order in which the spots of light he was seeing came on and then went out. "They didn't think I would be able to see a dark object, but light automobile or dark automobile, regardless what the color, the electrodes would still come on because we were in the sun. It was the same with people walking by. I could tell which way they were going," he said. As for how that made Churchey feel, he said, "I thank the Lord. That's all I can say."

When Churchey returned home, he took with him a camera mounted on a pair of eyeglasses; a video power pack and processor about the size of a videocassette, which he wears on a belt like a fanny pack; and the external transmitter, which magnetically clamps onto his skull over the implanted receiver-stimulator. The power pack has an on/off switch and a button that allows him to select among three strategies. With Strategy I, the electrodes stimulate the retina when the camera picks up light. Strategy II is a reverse mode; the processor fires the electrodes when Churchey points the camera toward dark objects. The third strategy is intended to respond to more distant objects than the first two, but Churchey said, he does not see much difference in that mode. For the most part, he keeps his system operating in the first strategy during the several times a day he uses it. Although he is well familiar with the whereabouts of everything in his home, he still finds it helpful to use the system, particularly when he wears it outside.

A highly religious man, Churchey told of sitting in his recreation room one day, his electrodes flashing on in response to light streaming though a window and thinking of the biblical phrase, " 'What hath God wrought?' I thought, 'how true,' " said Churchey. "After not being able to see anything in that eye, even though I just see spots of light, that's positive." In fact, he calls it "a blessing. I don't know how to put it any better way. From the first time they put that electrode in my eye down at Duke University in September of '92, I saw that one electrode, I knew they were on the right track. I didn't know when, where, or how, but I knew they were on the right track."

———◦———

THE OFFICES OF Second Sight, the company that built the retinal implants for Schoeman, Churchey, and other test subjects and that hopes to commercialize the technology, is on the right side of the tracks. It is situated in a nondescript two-story office building in Sylmar, about 25 miles northeast of Los Angeles, in an area populated by automobile repair shops and down-at-the-heels motels. But it is shielded from these establishments by a row of trees and a set of train tracks and is many light years away in its orientation. Once inside the doors of Second Sight, one finds the usual maze of offices and conference rooms, but there are also the much less usual clean room where electronic components for the implants are fabricated and laboratories where future generations of devices for the blind are being developed and tested. These operations are presided over by Robert Greenberg, the young president and CEO of Second Sight, who was a student of Eugene de Juan's when he and Mark Humayun were conducting their early neural prosthetic developmental work at Duke and Johns Hopkins universities.

Originally from Atlantic Beach, a small barrier island just off the coast of Long Island, and a stone's throw from New York City, Greenberg showed his affinity for both electronics and business at an early age. A computer geek, even before the word *geek* entered the vocabulary—at a time when personal computers were looming on the horizon—Greenberg developed software for several companies that were hoping to capitalize on the potential home computer market. Little did they know that Greenberg was still in his early teens studying electronics in a vocational high school program. But the future held bigger things for him.

After high school, he went on to Duke to study biomedical and electrical engineering on a full scholarship. As a member of the student government, he was on the buildings and grounds committee when he learned of a woman who had been raped after a dormitory door was left propped open. Turning his scientific and entrepreneurial bent to the problem, he spent the Christmas break of his junior year inventing an electronic device that prevented doors from being left open. It worked so well he was asked to build one for every dormitory on the Duke campus. His innate business sense saw a strong niche market, so he created a company to build and sell the systems, calling it Campus Security, and sent fliers to other universities around the country. Orders first trickled, then poured in. Greenberg began filling them by building the devices in his dormitory room, but as demand increased he had to outsource the fabrication of the system. Greenberg could have made this business his life's work, but he had other things in mind. "It seemed to me that I could do more with my electrical engineering training besides security systems. I had this idea that I wanted to apply electrical engineering to medicine," he said.

As his 1990 graduation from Duke approached, Greenberg felt that his training still had not given him the knowledge about human physiology he would need to pursue his interest in combining medicine and engineering. To fill that gap, he entered a combined M.D.-Ph.D. program at Johns Hopkins. Unbeknownst to Greenberg, he was following in the footsteps of de Juan and Humayun, who were both to be instrumental in his career path. He first met the ophthalmic team when he was two years into his studies at Johns Hopkins and they were preparing for their second experiment with Harold Churchey. Fascinated by this melding of medicine and electronics, Greenberg observed the test procedure, and just as Humayun had found the first experiment on Churchey a turning point in his professional life, so too did Greenberg find the second experiment a seminal experience. "It was really at that moment that I realized it would be possible to do a visual prosthesis; it was just going to take some engineering," said Greenberg.

Looking back at the experience years later, Greenberg acknowledged that as "a kid in medical school," his view of what it would take to bring the idea to fruition was a bit naive. "I thought it was going to be a piece of cake," he said. Yet he has never wavered from the feeling that "I knew it was possible at that point."

It was the excitement of this possibility that led Greenberg to go to work in de Juan's laboratory with an eye toward focusing his doctoral work on building a retinal prosthesis. He would have delved directly into designing such a device, but his program required that he conduct more basic research, so he turned to studying how electric current interacts with the retina. Two key findings of his dissertation research, which was conducted on frogs' retinas, was that the length of the electrical pulses introduced into the eye determines which cells are stimulated and that electrical stimulation of the retina causes greater activity in the bipolar cells than in the other layers responsible for processing visual information. This reinforced the idea that the retinas of people blinded by diseases affecting the rod and cone photoreceptors remain otherwise viable.

Greenberg completed his Ph.D. requirements in 1996 and his M.D. the following year. Convinced that the development of medical devices would be the direction of his career, he took a position with the FDA, where he hoped to gain an understanding of the regulatory process relating to medical devices. But he soon grew restless reviewing what others had developed. "One of the frustrations I had there was that I wasn't the one building the devices. It looked like everybody else was having all the fun, and I was getting to read about it," he said. To enter the fray, he became a product manager investigating advanced cochlear implants at the Alfred E. Mann Foundation, a nonprofit research organization in Valencia, California, devoted to the development of promising medical products. The foundation's goal is to nurture technologies whose anticipated payback is too far into the future to attract commercial investment and then license the technologies to commercial organizations once they are ready for the marketplace.

The foundation was established in 1985 by Alfred Mann, a scientist and entrepreneur. His early companies were Spectrolab and Heliotek, both based in Sylmar and involved in solar-powered photovoltaic systems. Mann then ventured into the medical device field by establishing Pacesetter Systems, Inc., one of the early cardiac pacemaker companies, which he sold to Siemens A.G. in 1985 and which was later sold to St. Jude Medical, Inc. Mann used the wealth he accumulated from these companies to establish the foundation that Greenberg joined.

One of the companies to benefit from the largess of the foundation was Advanced Bionics, also of Sylmar, which Mann founded in 1993. Advanced Bionics produces the Clarion, one of the leading cochlear implants. The technology for the Clarion evolved from research conducted at the University of California at San Francisco with NIH Neural Prosthesis Program support. Mann obtained the rights to the technology through a 1988 license agreement with the university. A group of scientists headed by Joseph Schulman brought that technology to fruition as the Clarion. When Greenberg joined the foundation, he went to work for Schulman and eventually ended up co-directing it with Schulman.

Alfred Mann was also instrumental in founding Second Sight, where he serves as chairman. But the idea for the company came from Sam Williams, a blind entrepreneur, who in 1955 started Williams International, a company that produces small gas turbine engines, and as its Web site says, is a "company with vision." Williams asked Mann if a cochlear-like implant could be created for the eye. Mann knew about the work of de Juan and Humayun, having met them several years earlier, and he was familiar with Greenberg's involvement with their research, so he had Williams speak to Greenberg about the idea. Following their conversation it didn't take Williams long to decide he wanted to invest in the concept. "Sam said that he had already given several million dollars to universities and nonprofit research foundations, and they generated some interesting science, but no product, and he wanted a product in his lifetime. He felt that the best way to do that would be to start a company and incentive the employees with options, the typical entrepreneurial approach. So he said he would only fund it if we started a company. Al agreed and that's how Second Sight got started," explained Greenberg.

The company was formed in 1998, and Greenberg was named president. Greenberg found it to be his "dream job. It was exactly what I had hoped to do." Unabashedly altruistic, Greenberg feels he can do more through Second Sight than if he practiced medicine, in which case he would be able to help only as many patients as he could personally see. But he feels if he can oversee the creation of a company and a device that hundreds of physicians implant, then he can have a substantial impact on thousands of people. "I found the ability to have a large impact on a large number of people very attractive, even though what I was giving up at the foundation was more hands-on research. There has always been the question of someday looking back over my life, and wondering, 'Will I have made a difference?' Maybe it's kind of a corny idea, but that's what really has driven me through all these years. I have this concept that we could do pretty much anything under the sun, so why not pick something that has the potential to make a difference?"

At the outset, Second Sight licensed the technology that de Juan and Humayun had been developing from Johns Hopkins, where the two were still located. Since Second Sight's sister company, Advanced Bionics, already manufactured cochlear implants, which were proven to be biocompatible and safe for human implantation, the retinal implant team felt the best way to start was to modify the cochlear implant equipment, enabling it to be used as part of the retinal implant. So with some minor alterations, the Clarion cochlear receiver-stimulator that is implanted behind the ear, became the receiver-stimulator for the retinal implant system worn by Churchey and Schoeman and known appropriately enough as the Model One retinal implant. Although this helped expedite the human testing process and saved millions of dollars in development costs, it still took two years to create the electrodes that contact the retina, as opposed to the two months that the scientists at Second Sight had hoped for. One of the major design challenges was creating an implant to attach to the retina that would withstand the eye's saccadic movements—continuous rapid movements of which we are not aware. "We were very worried about it," said Greenberg.

As Model One was being readied for human testing, work was already under way on the second-generation retinal implant. Instead of using the modified Clarion equipment, Model Two places all of the electronics inside the eyeball, which involves the monumental task of making the components approximately ten times smaller than they are in Model One. But the investment in time and money was deemed worthwhile since the newer system is safer, faster, and more economical to implant. And the Model Two implant comes equipped with sixty-four electrodes, a significant jump from the sixteen that Churchey and Schoeman wear. The purpose is to provide greater image resolution, much as more pixels in a digital camera provide a clearer photographic image. What kind of vision the greater number of electrodes will produce remains to be seen, but the goal is to give the wearer enough vision to allow reasonable mobility. When the new, improved system will be ready for human testing is uncertain, but once human testing is completed, Second Sight plans to commercialize the Model Two.

—⁄⁄⁄⁄⁄⁄⁄⁄⁄⁄⁄⁄—

TWO THOUSAND MILES FROM Second Sight, and on the opposite side of the retina, are the brothers Chow, Alan and Vincent, and their company Optobionics, of Naperville, Illinois, near Chicago. The Chows have created a completely different kind of retinal implant. The only two things the Optobionics chip has in common with Second Sight's system is that it sits on the retina, albeit on the opposite side, and that it is designed to return sight to the blind.

The Optobionics chip, called the artificial silicon retina (ASR), is implanted directly on the rods and cones located on the back of the retina. Thus, it is known as a subretinal implant. The Second Sight implant is attached to the front of the retina, where light first strikes before passing through the other layers to the rods and cones, so it is referred to as an epiretinal implant.

Alan Chow, a pediatric ophthalmologist, and Vincent, an electrical engineer, were born in Hong Kong and immigrated to the United States when Alan was 6 years old and Vincent was 8. Like many in the field of neural prostheses, Alan manifested interests in both medicine and engineering at an early age. While still in high school, his idea of melding a laser with a microscope to perform microsurgery won an award from NASA that earned him a trip to Cape Canaveral to watch the launching of Apollo 12. He did his undergraduate work at the University of Chicago and then attended the Stritch School of Medicine at Loyola University, also in Chicago. An abiding interest in microsurgery led him into ophthalmology, where dealing with the small components of the eye is an everyday affair. He decided to focus on pediatric ophthalmology because he found the prospect of helping youngsters, who still had many years to live, appealing.

The idea of developing a retinal implant had its inception in Chow's mind during the mid-1980s—around the time that Humayun conceived of his retinal implant—while he was a fellow in ophthalmic genetics at none other than Johns Hopkins University's Wilmer Eye Institute, a fertile ground for the growth of retinal implants. Though they each spent time at Wilmer, Chow did not meet de Juan or Humayun there, since he had moved on before they arrived from Duke.

While at Wilmer, Chow grew frustrated and saddened seeing numerous young patients with inherited eye diseases, such as retinitis pigmentosa, that he was powerless to do anything about. "All you really do is watch them get worse from year to year," he said. He was particularly struck by the case of a 10-year-old boy who was brought to him because it seemed as though he had gotten something in his eye while wrestling with his brother in the grass. Unfortunately, that was not the case. "When I examined him, his vision was 20/20 in both eyes and I saw absolutely nothing wrong with his retina," said Chow. "But over the course of three years, his vision deteriorated until it was about three times worse than legal blindness. I saw the typical effects of a condition called Stargardt's maculopathy, a disease similar to retinitis pigmentosa. I felt really frustrated that there was not too much I could do for him other than counsel his parents about career opportunities."

While contemplating what had not yet been tried to help young victims of such diseases, Chow struck upon the idea of restoring some vision through electrical stimulation of the surviving cells of the retina. He first raised the subject for discussion with Vincent at the Thanksgiving dinner table in 1988. Vincent immediately bought into the idea, and numerous discussions ensued during which the brothers considered the tack they would take. They eventually opted to build an implant to go on the underside of the retina, to put it directly in touch with the photoreceptors. They reasoned that since healthy rods and cones produce simple analog signals, by placing their implant directly against the damaged photoreceptors, they could provide the same less complex signals and let the remaining healthy cells of the retina do the complicated conversion to the digital signals required by the optic nerve.

In this regard, they were operating under the same premise as de Juan and Humayun, namely, that individuals blinded by retinal disease retain at least some viable cells in the other layers of the retina. The more layers of cells remaining viable, the better the implantee's vision would be, but even if only a relatively few cells could respond to the stimulation, some level of vision could be created. The Chows also felt that a subretinal configuration is preferable because it is naturally held in place between the retina and the back wall of the eye without having to be mechanically attached to the retina.

The brothers soon went about creating their implant and formed Optobionics in 1990 to help raise funds for the project. The resulting ASR is considerably less complex than Second Sight's epiretinal implant, in that it does not use an external camera or power source to provide input. Rather, it is, in essence, a miniature photovoltaic solar cell, the same as the increasingly popular devices used to convert sunlight to electricity for use in homes and offices. The Chows' solar cell for the eye is just over one-tenth of an inch in diameter (about the size of a pinhead), a thousandth of an inch thick, and includes some 5,000 microphotodiodes intended to provide pixels of light. It captures light and converts it to electricity to activate the retina. It is a simple yet controversial configuration; some experts say the ambient light that reaches the implant is not sufficient to provide the current needed to stimulate the retina.

One expert, who wished to remain anonymous said, "The system could not possibly work as they intended it to work. The amount of current produced by a photodiode is about a million times smaller per photon absorbed then the amount of ions, which flow around a photoreceptor. That means the sensitivity of a photodiode is a million times smaller than the sensitivity of a rod photoreceptor."

Alan Chow countered that because the implant is in the subretinal space, it needs much less current to stimulate the retina than an epiretinal implant. "In fact, we calculated approximately 10,000 times less," said Chow. "As a result, the implant can be entirely powered just by incident light without any batteries or external connections. The implant is totally self-contained within the silicon chip underneath the retina."

Twelve years after the Thanksgiving dinner conversation with his brother, Alan began experimentally implanting the chip in humans. The results have not been what he expected. Instead of the implant acting like a prosthesis, helping overcome a disability, in some cases it seems to be having a therapeutic effect, acting almost like a drug to reverse the damage to the retina caused by disease. Whether that actually is the case is also a matter of debate that remains to be resolved. Some experts say what the test subjects are experiencing is a placebo effect or the body's temporary response to the insult of surgery.

One of the strangest responses to being implanted was experienced by Ronnie Rainge, a physician himself, who suffers from retinitis pigmentosa and had an ASR implanted in his right eye on November 6, 2002. A 1976 graduate of Michigan State University Medical School, Rainge was serving in the U.S. Air Force as a family physician and was a few months from discharge when in 1980 he was diagnosed as having atypical sector retinitis pigmentosa, a rare nonhereditary type of the disease. He had been planning to leave the Air Force and go into emergency medicine, but the diagnosis changed all that. He knew he could lose his vision within a few years, or it could take longer. He decided to remain in the service, where he knew he would have financial security.

As it turned out, the progress of Rainge's disease was slow, which allowed him to continue to practice medicine at Air Force bases around the world for twenty years. Even when he retired on October 1, 1997, at the age of 50, he could still read and drive, though much of his peripheral vision was gone. Four years later, he started to lose the central vision in his right eye as well. When that happened, he moved from Huber Heights, Ohio, a suburb of Dayton, to Okemos, Michigan, near Lansing, to be closer to his sister Gloria Johnson, the person who learned of and urged Rainge to see Alan Chow.

Not one to sit back and do nothing, Gloria closely followed every bit of news she could find about retinitis pigmentosa, constantly looking for hope. In 1999, she read about the retinal implant work of Eugene de Juan and Mark Humayun at the Wilmer Eye Institute. She contacted their office on behalf of her brother and was told that an implant would not be ready for human testing for several years. Nonetheless, they asked that she send in her brother's medical records so they would have them for consideration when the time came. But Johnson did not find a willing partner in her brother, who was reluctant to provide his records because, said Johnson, "he just didn't think there was hope." Rainge said he felt that implant technology was "so far down the road I'd just wait and see how things were going to go." Johnson, on the other hand, did not want to sit tight. "I said, 'There is hope.' There are miracles, and I believe in them." And if she could help make them happen, so much the better.

Then Johnson read in an article that Alan Chow had already implanted three patients in June 2000 with the Optobionics device and was looking for more candidates. She was on the phone again, this time to Chow's office. Informed that there was already a long list of patients waiting for the implant, Johnson refused to take no for an answer. "I told them how much I love my brother, that he is a physician, and knowing how caring he was for others, I felt he deserved to be seen." Her persistence paid off; she was able to schedule an appointment for Ronnie, albeit for six months later. She needed that time to convince her brother to produce his medical records and to show up for the appointment. "I kept nagging him for his records. Finally I got them myself, and at three in the morning I went to Kinko's and made about ten copies of about one hundred pages of records," she said.

Rainge's resistance faltered in the face of his sister's persistence, and he agreed to meet with Chow. His first appointment was in December 2001. By May, there was no follow-up from Chow's office so Gloria was back on the phone, and in September Ronnie was in Chow's office for another exam. After five or six visits, he was accepted into the program and, said Gloria, "there was not a dry eye in the office that day."

By that time, Rainge had signed on 100 percent. He was convinced he wanted to receive the implant, not just for himself but for the benefit of others. He was fully aware of the risks involved but was undeterred. "To me, there was no question that I wanted to go ahead and have the procedure because there was so much to be gained. Just think of all the people who can benefit from this procedure. I realized this is only the initial phase of the study and knew that down the road, if this was successful I might get my sight restored, but if not, it would benefit those who come after. To me to be a part of that study, there's no way I would say I don't want to take any chances."

So on November 6, 2002, Ronnie Rainge underwent a two-hour surgical procedure at Central Dupage Hospital, near Chicago, during which an ASR was placed in his right eye. The response was quick and dramatic. He remained overnight in the hospital. The next day, he was discharged to his hotel suite, where he stayed with Gloria and her husband for two weeks.

At first, Rainge saw nothing through the implanted eye. Then, two nights after the surgery, he had a dream in which everything he saw was pink. He woke from the dream at about 1 in the morning. His right eye was covered, but with his left eye, the one that retained some vision and was not operated on, everything he saw in the room looked pink. "It kind of scared me. Was I having a hemorrhage into my eye?" wondered Rainge. "I wasn't having any pain or anything so I went back to sleep. I woke about a half-hour later and everything was back to the way it normally was. The color was gone."

He fell asleep again and awoke at 6:30 a.m. "I took the patch off and covered up my left eye to see if I could see anything out of my right eye, and I couldn't believe it," he said. "I had vision in my right. I could see things. I called my sister and said, 'I can see!' The vision in my right eye was almost as good as the vision in my left eye, and that was amazing because I previously had no central vision whatsoever in my right eye. I told my sister to hold up a finger and bend the finger and I told her what I was seeing. I could see myself in the mirror clearly. I was seeing out of my right eye for the first time in more than two years."

That vision lasted for an hour and a half, and as suddenly as it came, it disappeared. But Rainge was not distraught at having lost his vision for a second time. He was convinced it would return.

A few nights later, he had another dream in color, this time in black and blue, and just as the first time, when he awoke everything around him appeared to be black and blue through his left, "good" eye. This started happening on a regular basis. He would dream in either pink or black and blue and when he awoke everything would look as it did in his dream.

Equally strange was the fact that while nothing much was happening with his implanted right eye, the vision in his left eye was improving dramatically. Prior to the surgery, most of what he could make out through his left eye appeared either gray or black, but now "I could watch television and see the different colors on the football players' uniforms," said Rainge. He was also developing the ability to make out faces with his left eye and even see road signs, such as the yellow of the McDonald's golden arches. Meanwhile the color dreams continued, although less frequently.

Then, in May of 2003, six months after surgery, Rainge began to notice an improvement in the peripheral vision in his right eye. He could, for example, turn his head at an angle and see the burners on his stove and the vertical blinds in his kitchen. At the same time, the vision in his left eye continued to improve. He could even make out speed limit signs with his left eye as he was driven about town.

But a few months later, he was back on an emotional roller coaster as the vision in both eyes took a turn for the worse. A cataract started to form on his right eye, causing the vision he had realized since surgery to grow hazy. And the dramatic improvement he experienced in his left eye was reversing as well. "There are moments when I see things clearly out of my left eye. Not the colors, but the clarity is there. It comes and goes right now. I don't know what's going on," Rainge said at the time.

In March of 2004, he had surgery to remove the cataract. To add insult to injury, the ultrasound used to remove the cataract apparently vibrated the implant, leading to some hemorrhaging of the retina around it. This resolved itself with no discernable negative impact, but Rainge's vision in his implanted eye did not improve significantly even after he recovered from the cataract surgery. A year following implant surgery, the vision in his right eye reached a plateau of about 50 percent improvement over what he had been able to see before, and it has remained the same ever since. Before being implanted, he had some light perception in the right eye and could see motion but could not determine what was moving. Two years after surgery, he could see objects in his peripheral vision but still had no central vision. As for his left eye, it lost all the gains it had made and even became worse than it was before surgery, possibly because the disease continued to take its inevitable toll.

Some of the strange responses Rainge has had to the implant could be attributable to what Chow sees as the therapeutic effect the implant may be having. He has arrived at this conclusion based on the location of the chip on the retina. During the implant surgery, the chip is placed approximately 20 degrees away from the macula—the area of the retina that provides the sharpest vision and the most sensitive color vision. But instead of seeing the kind of images one would expect if the cells immediately under the prosthesis were stimulated, some patients, like Rainge, have experienced improvements in their vision that would seem to result from stimulation of the photoreceptors precisely in the macula region. This, said Chow, "is much more preferable to anything we could have anticipated or hoped for. We really didn't anticipate that this would be the case until patient after patient indicated improvement in the areas that were distant from the implants . . . that could not have been directly caused by the implant."

Chow's hypothesis is that once the retinal cells under the implant are stimulated, they produce neurotropic factors that signal other cells in the retina saying, "Hey guys, we're still around, wake up and start functioning again." The notion that the cells are dormant and not dead, and can therefore be awakened, is given credence by the fact that patients with retinitis pigmentosa do, on occasion, experience dramatically improved vision for short periods of time before relapsing. This is similar to muscle atrophy in individuals with paralysis. If the muscles are exercised, however, they seem to produce neurotropic factors that help keep them healthy. This is in keeping with the fact that the body, being a very efficient mechanism, shuts down what is not being used.

In the case of the retina, Chow's thinking is that when the other neurons in the retina, such as the bipolar cells, are not stimulated by the photoreceptor rods and cones, they do not produce the neurotropic factors to keep the photoreceptors going. This results in a vicious cycle. As the rods and cones lose their capacity to function properly, the other cells also produce fewer chemicals to keep them functioning. When the photoreceptors are stimulated, the negative process reverses and improvement occurs.

There are, however, skeptics in the field. One professor of bioengineering who asked not to be identified said, "It's been known for awhile, if I go into a person with very limited vision and do a procedure, if I inject saline in the vicinity of the retina . . . there will be a temporary restoration, a rescue of function. It has nothing to do with anything. It's just the act of that intervention that can produce a temporary rescue of function. We don't know why that is . . . It's not restoring vision, but it's a local rescue of function. It's possible it is due to a wound healing response." He compares it to the body responding to a cut or a splinter, mustering powerful mechanisms to restore itself to its original functioning state. And, he added, "there is also a placebo effect which is going on here too."

If Chow is correct, however, the question still remains as to whether his retinal implant merely staves off the eventual death of the retinal cells or actually reverses the degenerative process. Years after the first implants, all of the patients manifest some degree of visual improvement, according to Chow, so he has hope that the latter is true.

To determine the veracity of his theory, Chow received FDA approval to triple the number of experimental human implantees from ten to thirty. A significant aspect of the second phase of human testing is that the chips were to be implanted by physicians other than Chow, at medical facilities around the country, including Johns Hopkins Hospital, the Veterans Administration Hospital in Atlanta, the University of Washington, and Rush-Presbyterian-St. Luke's Medical Center in Chicago. Chow is also conducting ongoing animal studies to determine if the chip actually is slowing down or reversing degeneration of the retina's photoreceptors.

Despite the unanswered questions and the setbacks, Rainge said he is not discouraged. "I wasn't expecting to all of a sudden have 20/20 vision out of my right eye again," he said, adding hopefully, "Maybe with time it will improve." In the meantime, he feels he has bought time with the retinal implant; the improved peripheral vision he experiences with it has allowed him to continue to live alone and maintain his independence. "I am so happy to be a part of this study," he said. "Before, there was no hope. At least now there is hope. I think it's very promising. I really do. The next generation chip may be the answer to a lot of this. I'm hoping so anyway."

That hope cannot be underestimated. More than the increased physical ability, the hope and the sense of being part of something larger than themselves has given many retinal implantees an invaluable psychological boost that goes beyond the statistics of how well or poorly they can see. Rainge looks forward to his periodic follow-up visits with Alan Chow and the battery of tests he is subjected to: "It's so uplifting to have the tests performed and be able to see what you weren't able to see before. It boosts your spirits. It really does."

His sister attests to a dramatic difference in his spirits since the implant. "Before, he wanted to give up. He seemed very depressed," she said. "We all like to fish, to go boating, to travel. He would always say he wanted to stay home. But after he had the surgery, he ordered a special 26-foot pontoon houseboat made to his own design. He is a different person now. He has come out of that dark shadow."

Another of Chow's retinal implant patients, who has had a more consistently positive result than Rainge, is Maria Zaccaro of Rosemont, Illinois, near Chicago. An upbeat mother of two, Zaccaro was 42 years old when she received the visual prosthesis in 2002, just the day after Rainge was implanted.

Zaccaro first noticed a problem with her vision while pregnant with her daughter, seventeen years before receiving the implant. She could tell something was wrong, especially with her night vision, but it was hard for her to pinpoint exactly what. As time went on, the deterioration of her vision accelerated until eventually she could make out only light and dark, shadows and large objects. The diagnosis was retinitis pigmentosa.

When in 2000 she learned that Chow, who was her ophthalmologist, had developed a retinal implant and was looking for volunteers, she didn't hesitate to step forward. Thinking she had nothing to lose and everything to gain, she was unconcerned about having a foreign object placed in her eye. Her only fear was being put to sleep by an anesthetic, since this was to be her first surgery ever. As it turned out, the two-hour procedure at Chicago's Rush-Presbyterian-St. Luke's Medical Center to implant a chip in her right eye went smoothly. She stayed overnight and went home the next day. Zaccaro experienced no discomfort following surgery, the only side effect being a bloodshot eye that took about five months to completely clear up.

Within two months, she noticed an improvement in her right eye's vision. She found herself able to distinguish greater contrast between light and dark, rather than just whether a light was on or off. As time went on, she became able to make out specific objects, such as a cereal box, and eventually she could make out the large letters spelling out "Cheerios." "Before," she said, "I wasn't even able to see the light on a digital clock. I could see bigger lights like streetlights and lights on in a room, but I could not make out smaller lights on digital clocks, or alarms, or cell phones. Now I can make those out." Zaccaro has also regained some color perception. Prior to being implanted, she was able to discern the shape of a person. Now she can determine the color of the person's hair as well.

One of the key benefits of the implant is that she can now negotiate her way around a grocery or department store without having to be guided by her husband or one of her children. "I've always been very independent, wanting to do things as much as I can on my own, so I've learned throughout the years to do just about everything on my own," she said. "The only thing I don't do is drive or read small print like a book." As if to emphasize this point, there was a noise in the background as she spoke during a telephone conversation. She excused herself and returned a few minutes later explaining that a little girl she was babysitting for had to go to the bathroom.

"Babysitting?" she was asked somewhat incredulously. "Yes," she replied. She started doing so six months after she received the ASR, and said she would not be able to do so without it. Can she see well enough to handle scrapes and bruises? "Yes," she said. And are the parents who drop their children off at her house concerned about her limited vision? "Not at all," said Zaccaro. "They see that there has been much improvement. They have no question."

Questions about retinal implants still abound. Yet with cadres of researchers doing all they can to answer those questions, there is also hope.

Maria Zaccaro tells of receiving a telephone call from a retinitis pigmentosa patient after she appeared on a national television show to discuss her implant. He cried to her, "My God, there's hope for me."

"Yes there is. It's very promising," she replied.

11

Nerves of
Platinum and Iridium

The electrode is the business end, and it could be argued, the most important part of any neural prosthetic system. It is, in essence, the nerves of these devices, or more precisely that which replaces or bypasses nonfunctioning nerves. There is a wide range of electrodes that differ by function and type. There are those that record from nerves and others that inject electrical charges into them. They provide auditory, visual, or motor input. They are placed in the eye, the ear, the brain, in muscles, and around nerve bundles. And they come in a wide variety of sizes and shapes. There are probes that penetrate muscles and discs that sit on them and still others the size of grains of rice that are injected into them. Similarly, to interact with the brain, some electrodes are flat and are placed on its surface, while others are replete with arrays of minuscule pins that poke into the brain. Cochlear implants utilize tiny surface electrodes embedded in a coil that is snaked through the snail-shaped cochlea of the inner ear, while some auditory implants actually penetrate the brainstem.

Each of these configurations has advantages and disadvantages depending upon the specific application. And regardless of their configuration, electrodes must be able to efficiently, effectively, continuously, and safely do their jobs within the hostile saltwater environment inside the body over long periods of time. They must also be strong enough to withstand the rigors of implantation without damage yet malleable enough not to cause tissue damage during the same procedure. And they must be able to resist the body's natural instinct to reject all that it deems to be foreign matter. Developing such precise, delicate, and durable devices is a complex and tedious task not intended for those seeking quick results.

Many of the people involved in the development of neural prosthesis systems create their own rudimentary electrodes for their basic research. But when the need for electrodes that will stand the test of time arrives, they frequently turn to a small group of scientists who have devoted their entire careers to grappling with the thorny questions that must be answered before all the potential long-term benefits promised by neural prostheses can be fully realized.

One such scientist resides in a nondescript building adjacent to a multilevel parking garage across the street from Huntington Memorial Hospital in Pasadena. The building is home to the Huntington Medical Research Institutes, and the scientist is a tall, lean professorial man who sports a well-trimmed beard that has surprising little gray sprinkled in it for his 60-plus years. His name is Douglas B. McCreery, and his title, director of neural engineering, is, as is he, succinct, to the point, and somewhat understated. McCreery exudes consistency, even in his background.

Born and raised in the Hartford, Connecticut, suburb of Newington, he received bachelor's and master's of electrical engineering degrees from the University of Connecticut at Storrs in 1966 and 1970, respectively. He then went to work at the U.S. Naval Ordnance Laboratory at White Oak, Maryland, just outside Washington, D.C., where he focused on underwater acoustics. Yet his thoughts sometimes drifted back to his alma mater, where a doctoral program in bioengineering was taking shape just as he was leaving. This appealed to his interest in life sciences, which goes back to his kindergarten days, when he enjoyed "watering the plants and taking care of the bug colony." But when it came time to attend college, electrical engineering seemed a more reliable way to earn a living than life sciences, especially since he excelled in mathematics.

When bioengineering became a recognized discipline, however, McCreery decided to meld his engineering background with his love of life sciences and returned to the University of Connecticut to participate in the new program. His interest in neurophysiology began to blossom when, during his doctoral research, he investigated perception in catfish, specifically studying the lateral-line sensory system, which enables catfish to detect pressure waves underwater. This system is what allows fish to stay in formation while traveling in schools and is the evolutionary precursor to hearing in terrestrial mammals. Following the evolutionary chain led McCreery to investigate auditory physiology, to eventually help design the electrodes used in the House Ear Institute's auditory brainstem implant, and to lead the development of the penetrating auditory brainstem implant (see chapter 9).

After receiving his Ph.D. in 1973, McCreery did a postdoctoral fellowship in the Department of Neurosurgery at the University of Minnesota, where he studied the mechanisms of pain perception, focusing on how pain signals are modified as they travel from the spinal cord into the brainstem then to the thalamus before reaching consciousness. Much of this work explored these pathways in anesthetized animals. But McCreery was also able to study pain perception by using threshold stimuli in human subjects, most of whom were medical students who needed the $5 they received for allowing themselves to be human guinea pigs. During this period, the neurosurgical residents taught McCreery "basic survival neurosurgery."

While at Minnesota, McCreery also met Terry Hambrecht and Karl Frank, the founders of the then recently formed NIH Neural Prosthesis Program. Hambrecht had developed a neural prosthesis philosophy that he imparted to McCreery that left a lasting impression on the young researcher. It was Hambrecht's belief that neural prosthesis investigators needed to figure out ways to interact with the nervous system at its most basic organizational level, namely, with individual nerve cells or very small groups of neurons. Doing so would require the creation of microelectrodes that penetrate the nervous tissue to reach down into small clusters of neurons. This philosophy, said McCreery, "has been the linchpin of almost all of the work that I've done since I came to California" to join the Huntington Medical Research Institutes (HMRI) in 1979.

Even before McCreery arrived, HMRI was in the avant-garde of neural prosthetic electrode research. Its reputation was in large measure the result of work done by Robert Pudenz, a neuropathologist and neurosurgeon at HMRI, who during the early 1970s pioneered research into the mechanisms of tissue damage resulting from electrical stimulation. This work was carried on by William Agnew, who recruited McCreery as he was finishing his post-doctoral studies at Minnesota. Agnew also built on the work of Barry Brummer, an electrochemist at the Environmental Impact Center in Newton, Massachusetts (now EIC Laboratories), who was the first person to study the chemical reactions that take place during electrical stimulation of neurons with various types of metal electrodes.

What Brummer found put neural prosthetic scientists on notice. Within certain electrical current ranges, no significant by-products are produced, but beyond that the news was not good. Excessive levels of stimulation can lead to the creation of chlorine gas and hydrochloride, or bleach, which are toxic. Oxygen is also produced, which is not a problem unless it becomes gaseous and forms bubbles. Brummer based his research on passing current into a saline solution that simulated the fluid in the body. Pudenz took this work to the next level by conducting animal studies. In doing so, he found that the safe limits established by Brummer's saline solution studies were indeed accurate and that these limits are essentially the same throughout the body.

In his studies, Brummer also found that the by-products produced by electrical stimulation differed greatly depending upon the type of metal used in the electrodes. He initially studied pure platinum, the most commonly used material for such applications. He also examined the use of silver, which was a popular electrode material during the early days of neural prosthetic development, and he found that it created a high level of toxicity, as does copper. As a result of this work, these metals are no longer used in prosthetic electrodes. When McCreery arrived at HMRI, he picked up where Pudenz, who had recently retired, left off.

THE FLOW OF ELECTRICITY CAN BE thought of as waves of water, with the amplitude being the difference between the top and bottom of the waves and the current being the rate of the flow of the charge (electrons or ions). If electricity is injected at either too high an amplitude or too high a rate of flow, damage can be done to the nerves receiving the charge. Too high an amplitude, for example, can electrocute nerve tissue in a process called electroporation. Overstimulation can also lead to a buildup of acid or alkalinity around the electrodes, which in turn causes toxic elements to be introduced into the body in a process called asymmetrical oxidation and reduction. The word *asymmetrical* is used because the cause of the problem is an unequal introduction of negative and positive charges into the body. The cure for this condition is to make sure the stimulus is charge balanced—that equal amounts of positive and negative current are fed to the electrodes.

Based on their extensive research in the field, the HMRI scientists formulated "prudent electrical stimulation" guidelines for developers of neural prostheses. This three-step process consists of the use of a noble metal, such as platinum or iridium; the use of charge balance stimuli, meaning injecting equal positive and negative charges; and the use of an appropriate stimulus pulse rate and amplitude.

While McCreery and his colleagues labored over these difficult and complex issues, he was approached by the House Ear Institute for assistance in developing electrodes for the auditory brainstem implant (ABI). The institute had initially developed its own electrodes designed to sit on the surface of the cochlear nucleus in the auditory brainstem, but when they were implanted in a patient and turned on, she experienced twitching and tingling in one of her legs. The House researchers thought her reaction might be caused by movement of the electrode following implantation of the rather crude electrodes they were using, and they turned to HMRI for help in designing a better electrode.

Over a period of two years, McCreery and his group worked to develop a second-generation ABI electrode that utilizes the body's tendency to create scar tissue around any implanted object to hold the implant in place. The resulting design was successful at helping to anchor the electrodes, but it still utilized electrodes placed on the surface of the auditory brainstem. Not long after the development of that electrode, Hambrecht challenged

McCreery to pursue the goal of addressing small groups of individual nerves by designing microelectrodes that would actually penetrate into the auditory brainstem.

Driven by the researchers at House, who were hoping to improve the performance of the ABI, and Hambrecht having laid down the gauntlet, McCreery took up the challenge. But not without some trepidation. "My God," he recalled telling Hambrecht, "we've got to start at square one. We know nothing about how to implant these electrodes safely, how to push them into the brain with minimum mechanical damage, how to configure the tips to minimize the damage. We've got to invent a whole new technology for this." McCreery remembered Hambrecht's cool response to all of this being one word: "OK." Shortly thereafter, HMRI received an NIH contract from Hambrecht's program to proceed with work on a penetrating electrode.

McCreery first tried using the same platinum-iridium alloy he used for the surface ABI electrode but soon realized he would have to change course. Though platinum and iridium alloys are fine materials for surface electrodes, they are less than desirable metals for penetrating electrodes that have sharp points, which like lightning rods concentrate the electrical current to a higher level than what is delivered by the larger surface electrodes. The tip of a penetrating electrode looks like a speck of dust to someone with keen vision, while an ABI surface electrode resembles a decorative sparkle.

The problem caused by increased current concentration from the smaller electrode arises from the battle between electrons and ions when electricity is introduced into the body via metallic electrodes. The battlefield is the precise point where electrode meets biological tissue. That is where electrons and ions clash. The entity that carries an electrical charge in metal—the electron—is different from the charge carrier in biological tissue—the ion. An electron is the basic unit of charge in the universe. An ion is an atom consisting of a nucleus containing protons, which are electrically positive, surrounded by a cloud of electrons, which are negative. When an atom has an equal number of protons and electrons, it has no charge. If an atom loses or gains an electron, it has an electrical charge and is by definition an ion.

Unfortunately for the developers of neural prostheses, electrons do not pass through solutions—human tissue being primarily fluid—and ions do not flow through metal. So when an electron reaches the tip of an implanted electrode, it cannot flow through the surrounding tissue. As a result, electrons tend to pile up at the tip of the electrode and ions pile up in the tissue adjacent to the electrode, causing the interface to become more and more highly charged. It is at this interface that a lot of the toxicity problems are generated. During the battle between the electrons and ions, electrons leap to the ions neutralizing them, with the result that an electron current flows through the metal electrode and an ionic current flows through the tissue. The goal of getting an electrical charge into the tissue is accomplished, but at a price. Essentially, corrosion of the metal—called an oxidation-reduction reaction—takes place, which if not properly controlled, can cause toxic materials to enter the body.

While searching for an acceptable material for the PABI electrode, Mc-Creery learned that Lois Robblee, a colleague of Barry Brummer at EIC Laboratories was testing activated iridium as a possible neural prosthetic electrode material. It turns out that iridium with iridium oxide deposited on the surface can intercede in the battle between ions and electrons at the interface, thereby allowing increased stimulus levels before the corrosion of the iridium metal electrode occurs. The iridium oxide reduces corrosion by first increasing its oxide coat and then when the current is reversed, allowing some of its oxide coat to be removed by a safe oxidation-reduction reaction. The net result is that the electrode remains stable. This allows higher electrical charges to be safely injected into neural tissue than if either platinum or pure iridium were used. Activated iridium has the added advantage of being a hard material, which is desirable for electrode insertion into neural tissue. McCreery and his team joined forces with Brummer and Robblee to develop methods of shaping the material into microelectrodes.

The difficulty of designing electrodes to implant in the brain is made worse by the fact that the brain contains a dense meshwork of blood vessels and "bleeds like hell if you cause any insult to it," said McCreery. Getting around this problem presents several conundrums. First, the electrodes must be shaped to push the blood vessels aside as they enter the brain. If the tip is not sharp enough, it can cause extensive tissue damage when inserted, but if it is too sharp, the density of the current at its tip becomes too great and increases the chances of corrosion. Then there is the question of electrode insertion, since those who design electrodes must also make the tools with which they can be implanted. If the electrodes are inserted too slowly, the tissue tends to tear, yet inserting them too rapidly can cause other types of injuries. Add to this the fact that those developing the insertion technique and instrumentation must make sure the electrodes themselves are not damaged during insertion and the complexity of the entire process becomes clear. So it is little surprise that it took several years to create a viable penetrating electrode array and insertion device for the PABI.

The resulting array contains eight needlelike electrodes, each less than one-hundredth of an inch in diameter at its base and even smaller at the tip. The electrodes are of varying lengths—up to a maximum of 2 millimeters (seven-hundredths of an inch)—to reach into various strata of the cochlear nucleus. The paddle on which they are mounted resembles a minuscule hair brush with progressively smaller bristles. The insertion tool is a small hand-held device that includes a gunlike trigger, but instead of bullets it fires electrode arrays into the cochlea nucleus at about 1 to 2 meters (about 6.5 feet) per second. Not inconsequentially, the electrode array must be produced painstakingly by hand under a microscope.

The successful development of the penetrating ABI electrode laid the groundwork for McCreery and his HMRI colleagues to launch into the creation of penetrating electrodes for a visual cortex implant undergoing development at the Illinois Institute of Technology in Chicago. Researchers there estimate that their prosthesis for the blind will require at least one thousand electrodes, and as McCreery said in his typically understated way, "that is a lot of electrodes." The reason for the vast difference between the PABI's eight penetrating electrodes and the estimated one thousand for a visual implant is the considerably greater amount of information presumed to be required by the brain to create a visual image than is needed to process sound.

To create such an implant, the HMRI experts are taking a tiling approach in which they stack small electrode arrays side by side to give the required total. This raises the question of how the visual cortex would react to having as many as one hundred arrays of ten electrodes each penetrating its surface in close proximity to each other. Regardless of how well designed they are, there is always a certain amount of trauma to the tissue when probes penetrate the brain.

An obvious way to optimize the number of electrodes while minimizing neural damage is to shrink electrodes down to sizes even smaller than they already are, but doing so would make it impossible to produce them by hand as they have been. To make components still smaller requires utilizing the same photolithographic processes used to place millions of transistors on minuscule semiconductor chips, which when applied to neural prosthetic electrodes presents its own series of complex problems. Yet bringing semiconductor-production technology to the world of neural prostheses is precisely what a number of researchers are doing.

One of the leaders in the field is Kensall D. Wise of the University of Michigan, who has been referred to as the father of solid-state prostheses. A professor of electrical engineering, computer science, and biomedical engineering and director of the Engineering Research Center for Wireless Integrated Microsystems, Wise has been tackling the intricate problems involved in adapting solid-state technology to fabricate electrodes that interface with the nervous system since 1966. At the time, he was an electrical engineering graduate student at Stanford University in search of a dissertation topic. Fortuitously, the electrical engineering department was promoting joint projects with the medical school. This led an engineering professor to propose applying a new semiconductor technology being developed at Bell Laboratories for nonneural purposes, to produce microelectrodes to record from neurons.

Known as beam lead technology, the new concept involved producing silicon chips that had gold leads sticking out beyond their edges so they could interconnect with other electronic elements to create integrated circuits. The Stanford faculty member thought the same technology could be used to create implantable devices with gold tabs that would serve as electrodes to record electrical signals from firing neurons. He suggested recording from neurons rather than stimulating them, because recording was, and still is, a commonly used tool for studying the central nervous system and the focus at Stanford was on providing a better set of tools for neuroscience.

Wise, who received a master's degree in electrical engineering from Stanford two years prior, had just returned to Stanford from Bell Labs in Murray Hill, New Jersey, to work toward a doctoral degree. With his Bell Labs background, he felt the subject was a natural for him to pursue. When he started working on his dissertation project, Wise's primary interest was integrated circuits and not the biomedical application of electronics. Yet it wasn't long before he found the lure of applying electronics to biomedicine irresistible. Coincidentally, a senior faculty member in Wise's department had a daughter who was blind. In an early application of image sensor technology, the father invented a device that could read text aloud to his daughter. Though Wise was not involved in the project, the young, blind woman made a lasting impression on him, and years later he speculated that she may well have had an influence on his decision to dedicate his life's work to the development of electrodes for neural prostheses.

But the road he took was not direct. He received his Ph.D. in 1969 and remained at Stanford to continue the beam lead work as a postdoctoral fellow until 1972, when he went back to Bell Labs and turned his attention to things other than neural prostheses. Two years later, he moved to the University of Michigan, where a relatively small portion of his research was devoted to neural prostheses. That changed dramatically in 1981, when he procured a contract from the NIH Neural Prosthesis Program to delve deeply into the creation of solid-state electrodes for both stimulating and recording from the nervous system.

A driving force behind Wise's return to the neural prosthetic fold full-time was the great change that occurred in electronic photolithography between 1972 and 1981, making the notion of solid-state electrode production considerably more viable. It had previously been something of a hit-or-miss chemical operation, and the standardized production of high-quality electrodes was a virtual impossibility. Now, improvements in the "etching" phase of photolithography presented the realistic possibility that automated, high-yield production methods could be applied to consistently manufacture first-rate electrodes.

In the photolithographic process, layers of material are placed on a silicon substrate and the desired features are photographed onto the chip using a mask, comparable to a photograph negative. Then, just as chemicals are used to print a picture on paper, the photolithographic process uses chemicals to etch away the undesired portion of the chip, leaving the required functional areas intact. Though it sounds relatively easy, there are approximately one hundred complicated patterning procedures involved in using this technology to form electrodes. Nonetheless, thanks to the recent advancements, within four years, Wise was able to develop an electrode production process that has not changed significantly since.

But creating consistently good-quality electrodes was far from the end of the battle. Since Wise resolved the production problems, he has devoted himself to answering a number of questions that those creating chips for conventional electronic devices do not have to deal with. Primary among them is whether the many layers of material laid down on the silicon substrate will delaminate over time inside the human body. Another issue is that silicon wafers are essentially glass, which means they can break. Designing them so they are small enough to be implanted into human tissue, while retaining the necessary flexibility and strength so they don't break has been a major challenge.

Meanwhile, Wise and his group have also been busy developing different types of recording and stimulating probes for a wide range of applications and have since created over one hundred different probe designs. In the process, his laboratory has become an electrode supermarket for neural prosthetic researchers, who can obtain off-the-shelf electrodes or have them custom made. This concept became so successful that in the early 1990s the NIH funded the creation of the Center for Neural Communication Technology at the University of Michigan to give Wise the wherewithal to continue providing electrodes to other researchers around the world.

Wise has also been pursuing the mantra of virtually all electronic component designers: smaller. His specific quest has been for smaller arrays on which to include more microelectrodes, bringing ever closer to achievement Hambrecht's goal of addressing individual neurons. As a step in that direction, Wise's group has used their semiconductor acumen to create a 1,024-electrode implant resembling a miniature pitchfork. To communicate with these probes once embedded in the brain, they have included hardware that can process, amplify, and even multiplex electrical impulses using wireless telemetry, all built right into the probe.

The entire apparatus is about the size of a small button and eliminates the need for any wires running from the probe to a separate receiver-stimulator. In the multiplexing mode, the system can receive impulses from and send signals to all of the electrodes in such rapid-fire succession that it seems like a continuous flow of information. Advanced as this system is, it is still huge by modern microelectronic standards. As Wise sees it, the only thing still needed to build solid-state neural prosthetic electrode systems with one hundred times more functionality in the same space is, in a word, money.

In the interest of expanding applications for its solid-state technology, Wise's laboratory is also working on dual-purpose electrodes that can deliver drugs as well as electrical stimulation. An early application will be as a research tool to study the effects of certain drugs at the cellular level. Eventually, this may provide new ways of treating neurological disorders such as epilepsy and Parkinson's disease with a combination of electrical stimulation and chemical control. "If we implant a button-size electrode array in an epileptic focus, a place where seizures begin, and the array listens to activity and detects the beginnings of a seizure, it can either use electrical stimulation to change the patterns of activity, or it can medicate the tissue by delivering drugs locally to ensure that it doesn't develop into a full-blown seizure," said Wise. The expectation is that the drugs will be supplied to the implant by a reservoir implanted under the skin close enough to the surface that it can be refilled periodically.

RICHARD A. NORMANN, ANOTHER LUMINARY in the field of neural prosthetic electrode development, traces his interest in things electrical to pinball machines. When he was a child in the San Francisco Bay area, Normann's father, a Danish immigrant who dreamed of owning a farm in the United States, bought and installed pinball machines to raise enough money to purchase a farm, which he eventually did. In the interim, the Normann basement became a pinball aficionado's dream. The collection of old and broken machines gave the young Richard plenty of opportunity to tinker, which as a self-admitted nerdy type, he was happy doing. From working on those broken-down pinball machines, Normann has since risen to head the Center for Neural Interfaces (CNI) at the University of Utah, where he is also a professor of bioengineering, ophthalmology, and physiology.

Naturally enough, when it came time to attend college, Normann majored in electrical engineering at the University of California at Berkeley, where he earned bachelor's, master's, and doctoral degrees. During his student days, he became interested in merging engineering with biological sciences but found, as did many other pioneers in the field of neural prostheses, that bioengineering did not exist as a formal discipline. Undaunted, Normann became something of a self-made bioengineer by crafting his own curriculum, which included courses in physiology, neuroscience, and anatomy. He incorporated his dual interests into his Ph.D. thesis, which dealt with how the rods and cones of the vertebrate retina adjust to light intensity so that we see well whether on a ski slope that is showered in intense sunlight or bathed in moonlight.

Following receipt of his doctorate in 1973, Normann took the path of so many others interested in bioengineering and joined the National Institutes of Health, where he continued basic research on the workings of the retina. Five years later, he moved to the University of Utah, which by then had a department of bioengineering and where he continued to follow the path of retinal research. He would have been quite happy to stay on the same career track, save for the fact that he began to find it difficult to attract bioengineering students to work in his laboratory on the basic science questions he was pursuing. To resolve the problem, he decided his research needed some sex appeal, or in his words "an applied spin," to the subject matter. So he struck upon the idea of artificially creating vision in the blind by electrically stimulating the visual cortex. He knew that there had been interest on the part of other scientists in doing so but there seemed to be little progress since Giles

Brindley implanted several patients over fifteen years earlier, and Normann wondered why. To find out, he invited a group of distinguished researchers in the field of neural prostheses to a meeting at his home in 1982, and he posed that question to them.

"The consensus of that meeting was that artificial vision had not moved, that neural prostheses had not moved very effectively, because nobody had come up with an appropriate electrode array architecture that would really enable scientists to talk to large numbers of neurons with good selectivity," said Normann, who resolved to solve the problem.

Taking a visual cortex approach to return sight to the blind seems like an odd direction for someone who had spent many years studying the retina. But Normann said it was specifically because of his experience working on the retina that he chose to take the visual cortex route. "I knew the retina is an extremely delicate piece of tissue," he said. "A cortical approach would seem to be a more rational approach. It's a much more robust environment, the cortex is protected by the skull, and does not have the high accelerations associated with saccadic eye motions. And the cortex is also spared the extensive degenerations of many blindnesses, such as those caused by trauma to the eyes or the optic nerve experienced through gunshot wounds or automobile accidents. So the cortex looked like a much more appropriate place to start stimulating."

Normann also opted to follow Terry Hambrecht's precept of attempting to activate small groups of individual neurons by creating an implant with needlelike points that would penetrate the cortex rather than sit on its surface —a route some others in the field were taking. In addition to offering better selectivity, stimulating neurons with penetrating electrodes requires considerably less current than do electrodes that sit on the surface of the brain. And the larger charges required by surface electrodes have been known to trigger epileptic seizures.

Now that Normann knew that he wanted to stimulate the visual cortex with penetrating electrodes, two more requirements remained to be fulfilled before he could develop a system. He still needed to find graduate students to help with his research, which is what led him to this project to begin with, and he needed funding. Help came to him serendipitously in the form of a young man named Patrick Campbell. In a stroke of perfect timing, Campbell, who had until recently been working on designing a cochlear prosthesis for a local company, walked into Normann's office and told him that he wanted to return to school to obtain a doctorate in bioengineering while continuing work on neural prostheses. Normann and Campbell became a team, and over the next six months, "we begged, borrowed, and stole instrumentation" to begin work, said Normann. With a concrete idea of how to proceed, a worthy graduate student, and a laboratory ready to go, Normann was now able to obtain seed money from the National Science Foundation to proceed with his project.

It took three years or so to create the Utah electrode array, which is made of silicon, glass, and platinum and is produced using a blend of old-fashioned handwork and high-tech photolithography. The result is a 4 x 4–millimeter (0.16 inch) array of one hundred electrodes that looks like a bed of nails that a small ant might lie on, with each electrode being 1.5 millimeters (0.06 inch) long.

Even though Normann envisioned his electrodes being used to stimulate the visual cortices of people who are blind, its first human application was to record from the motor cortex of a man with paralysis. The change in direction from a stimulating to a recording electrode came about when Normann met John P. Donoghue, inventor of the BrainGate brain-machine interface system, designed to record the firings of neurons in the motor cortex and use them to enable a patient to operate a computer. Donoghue was in search of electrodes to use in conjunction with his system and was impressed with Normann's Utah electrode array, which was implanted in a human for the first time in 2004 (see chapter 12).

Normann has since developed what he calls a slanted Utah electrode array, which consists of progressive rows of ever shrinking electrodes and looks like a slanted bed of nails. It is designed for use in conjunction with motor prosthetic systems to activate the limbs of individuals with paralysis by stimulating peripheral nerves as they branch throughout the body. These nerves travel in bundles known as fascicles, in a configuration similar to a telephone cable that carries numerous wires branching off into smaller bundles of wires until it gets down to the two wires that provide service to your home. So too do neurons branch off until the individual fibers reach the muscle they are destined to stimulate. The idea behind the slanted array is to access a range of neurons at various depths within a fascicle.

—◁◁◁◁◁◁▷▷▷▷▷—

THE MOST UNUSUAL AND FUTURISTIC of the family of neural prosthetic electrodes being developed is the BION, as in bionic, the linking of *biology* and *electronics*. Not much larger than a grain of rice, the BION is designed to be a freestanding electrode, no wires needed, that will stimulate paralyzed or weakened muscles. Because of its small size and versatility, the BION can theoretically be implanted anyplace in the body where stimulation is needed, as in the tongue to alleviate sleep apnea.

Like a ship in a bottle, the BION is a cylindrical piece of glass or ceramic that contains the electrode itself as well as signal-processing electronics and in some cases its own built-in power supply. Those BIONs that do not contain rechargeable batteries receive their power along with operating instructions by means of radio frequency signals, and they include the built-in circuitry to convert those signals to power, stimulating instructions, and stimulation pulses. A single external coil can address many implanted BIONs. These miniature devices offer a number of potential benefits over the more conventional FES electrodes, which are frequently larger and are usually connected to microprocessors implanted elsewhere in the body through a web of embedded wires.

By eliminating the need for these wires with their own built-in microprocessors, BIONs reduce pathways for potential infections and eliminate problems with lead wire breakage. And BIONs can be injected into muscles in a less invasive procedure than is required for the insertion of other neural prosthetic electrodes. The BION implant process can be performed in a doctor's office using an insertion tool that is a modified version of an instrument commonly used to start intravenous lines. The skin is anesthetized, and a tiny puncture is made with a scalpel to make it easy to push a needle through the skin and into the muscle. An electric current is run through the needle, and the resulting muscle contractions are observed to determine the optimal location for the BION. The needle, which is surrounded by a plastic sheath, is then withdrawn, and the BION is slid down the sheath to the desired location, and the sheath is removed. Total implantation time runs about twenty minutes, and the wound generally heals within a few days.

BION-based systems also have the advantage of being easily scalable. A BION neural prosthetic system can conceivably consist of one or hundreds of BIONs implanted throughout the body, depending upon the intended application. "You don't have a fixed, complex, single device that all patients have to be treated with," said BION inventor Gerald E. Loeb. "You have a set of modules. You can put them in, see how it goes, put in other modules if it's going well or if you need a different function, and work with the patient, gradually building up whenever there is a need." One caveat is that coordinating a large number of BIONs to fire off at just the right time to create smooth synchronized movement is a major challenge.

Work on BION development began during the late 1980s and has taken some time to reach the human testing stage for two basic reasons. First was the difficulty involved in figuring out how to create something as small as a BION, with all of the required advanced electronics being self-contained. The other problem revolved around hermetically sealing these minuscule devices so they could be implanted in the hostile environment of the human body and remain viable for decades. These problems have largely been addressed by two BION development projects, each answering these questions in its own way and each having branched off from the same family tree.

One BION effort is sponsored by the Alfred E. Mann Foundation of Valencia, California, whose president and chief scientist is Joseph Schulman. The second BION project is being conducted under the auspices of the closely related Alfred E. Mann Institute for Biomedical Engineering, based at the University of Southern California. The institute's BION effort is headed by Jerry Loeb, a physician and director of the Medical Device Development Facility of the institute. Loeb is also a professor of biomedical engineering at USC and the deputy director of the Center for Biomimetic Microelectronic Systems, headed by retinal implant developer Mark Humayun.

Loeb's interest in neural prostheses dates back to 1966, when he was a student at Johns Hopkins University in a combined undergraduate and medical school program. During the summer between his freshman and sophomore years, he got a job at the Johns Hopkins Applied Physics Laboratory. Microelectronics was in its infancy, and the laboratory was one of the few places developing the technology. Loeb was assigned to a project aimed at engineering microcircuits to talk to neurons, and he ended up building one of the first microelectronic arrays to record the electrical activity of living cells. Specifically, he created a tissue culture array in which electrodes were printed on glass and cells were grown on top of them. This basic research project served as the foundation of Loeb's future professional path that led him into the world of neural prostheses. And, coincidentally, the interplay of glass and electrodes continued to have a large role in Loeb's professional life, since he is currently pursuing glass BIONs, while Schulman is working on building ceramic BIONs.

A few years after Loeb's first foray into developing electronics to interact with neurons, a Johns Hopkins neurosurgeon received an NIH contract to fabricate and test electrode arrays for a visual cortex prosthesis being developed at the NIH. Because Loeb, who was by then a medical student, had experience in the area, he was recruited for the project, where he helped design electronic equipment associated with neural prostheses, such as amplifiers and microprocessors, in addition to working with electrodes. He was also involved in animal tests, and even some early human experiments in which the visual cortex was stimulated while patients underwent thalamotomy surgeries, which involve destruction of part of the thalamus—a part of the brain involved in processing sensory input—to alleviate chronic pain. During these procedures, which are performed under local anesthesia, the researchers placed stimulating electrodes on the shank of the probe that was to eliminate the appropriate part of the thalamus. As the probe passed through the visual cortex, the electrodes were activated and the patients were able to tell the investigators whether they saw anything.

While working on the NIH visual electrode endeavor, Loeb met William Dobelle, who was pursuing a master's degree in biophysics at Johns Hopkins and was also a student investigator on the project. Subsequently, Dobelle moved to the University of Utah, where he set about creating his own visual implant system based on electrodes that sit on the visual cortex of the brain. Dobelle invited Loeb to join him at Utah to help with the electrode development. Though still a medical student, Loeb took a six-month elective in 1971 to join Dobelle, and while there, he designed the electrodes for Dobelle's system. The NIH began to back away from the surface electrode approach during the mid-1970s, primarily because the power required to induce spots of light in a patient's visual field was deemed too high. The current could spread to other parts of the brain and cause epileptic seizures, which did happen on occasion. Nonetheless, Dobelle went on to become a controversial character in the field of neural prostheses, as he implanted electrodes on the visual cortices of blind individuals with mixed results. Those were the days before the FDA regulated such implants. Later, when Dobelle decided to commercialize his visual prosthesis, he moved his operation to Portugal.

After graduating from medical school in 1972, Loeb spent a year as a surgical resident at the University of Arizona. He then followed what was becoming a well-worn trail for those interested in neural prostheses to the National Institutes of Health, where he worked under Terry Hambrecht. Loeb remained at the NIH for fifteen years, eventually heading his own research group that worked on developing electrodes capable of recording from sensory cells in the muscles and skin (proprioceptor and somatosensory receptors, respectively) that provide information to the brain and spinal cord for movement control. These receptors feed data to the brain relative to muscle position, force, stretch, and velocity at very high rates. That is what enables you to move your arm around with your eyes closed and still know its location.

Recording from these sensors, thanks in part to Loeb's early work, has added to the knowledge of how the brain incorporates that data and uses it to know which muscles to fire off and what the appropriate tension should be. Yet the ability to record from touch and muscle sensors is still well behind the art of using electrodes to record neural activity in the brain. As a result, one of the unanswered questions of brain-computer interface, and functional electrical stimulation, is how well firings from the brain can direct implanted prostheses without normal proprioceptive and somatosensory feedback.

Experts suspect that it is highly unlikely that anything approaching natural movement will be achieved until such feedback can be provided to neural prosthetic systems. It is one thing to record brain signals from the motor cortex and have a computer analyze and transpose those signals into jolts of electricity that activate a human or robotic arm. It is a completely different, and infinitely more complex, thing to calculate exactly how much force will be needed in all the muscles to produce the exact trajectory desired and the amount of stiffness required to accommodate unexpected loads or disturbances, as when an arm bangs against a chair. This complexity is made apparent by the fact that the nervous system lavishes much more attention on the sensory side than on the motor side. Over ten times as much information is transmitted from the muscles to the spinal cord and brain than there is data going from the spinal cord out to the muscles.

After spending three years or so working on recording from sensory cells, Loeb's focus shifted when he was named a project officer in the Neural Prosthesis Program's effort to promote the development of cochlear implant electrode arrays. Though the project officer's job is usually that of an administrator, Loeb became particularly interested in and involved with developmental work being conducted at the University of California at San Francisco, where he participated in the creation of a four-channel electrode array that was the forerunner to the Clarion cochlear implant, one of the leading commercial cochlear implants.

All of this was prelude to Loeb's idea for the BION, which he hit upon while consulting for a company that designed electronic dog tags to be injected into animals for identification purposes. The company had created the operational part of the tag and asked Loeb to devise a hermetically sealed capsule that could contain the electronics, could be injected into an animal, and would remain functional in its body on a long-term basis. This led Loeb to think about using a similar system to stimulate paralyzed muscles, which he proposed to Terry Hambrecht. As it happened, Hambrecht and his deputy at the Neural Prosthesis Program, William Heetderks, had been seeking out technology that would enable electrodes to function in a wireless mode. The BION idea had the potential to provide a platform for developing just such wireless communication technology while also being able to serve as a single-channel implantable stimulator in its own right.

But there was a fly in the ointment. As Heetderks explained it, "When Terry and I first started this development, there was considerable skepticism as to whether it would be feasible. In particular there was concern about the amount of power that would be required to run such a system. I remember one person commenting that it might work fine as long as the person was willing to stay within 20 feet of a power substation."

To test the concept before backing it with NIH funding, Heetderks built a demonstration system, which proved that a self-contained microstimulator, such as the BION, could stimulate neurons using only the power derived from a radio frequency field. He subsequently wrote a paper on the demonstration entitled "RF Powering of Millimeter- and Submillimeter-Sized Prosthetic Implants," which appeared in the May 1988 edition of the IEEE Transactions on Bio-Medical Engineering. "It is fair to say that the project would not have gotten started without Jerry's [Loeb's] suggestion, without Terry's [Hambrecht's] encouragement and support, without the feasibility study, or without the potential additional application to the microelectrode arrays— not to mention the support of the taxpayers who ultimately paid the bills," said Heetderks.

Having recently left the NIH for Queen's University, in Kingston, Ontario, where he became a professor of physiology and joined the university's Biomedical Engineering Unit, Loeb was free to bid for funding that the NIH was now making available for BION development. In the end, the funds were divvied up among three institutions, each designated to work on one part of the creation of a BION. Having worked on a capsule for the animal implant, Loeb was made responsible for the packaging. Another group, based at the Illinois Institute of Technology, was to design the external control system that would transmit instructions and power to the implanted BIONs. That group was headed by Philip R. Troyk, who currently leads a team focusing on visual cortex implant technology. The third member of the BION triumvirate was the Alfred E. Mann Foundation, run by Joseph Schulman. Schulman's group took on the task of creating the electronics to go inside the BION. The Mann Foundation was also designated the repository for any intellectual property that came out of the project for technology transfer and commercial licensing purposes. Since then, some BION technology has been licensed to Advanced Bionics, of which Al Mann is the chairman and co-CEO.

Building on the technology he developed for the animal transponders, Loeb decided to develop a glass capsule for the BION, but he had difficulty solving a leakage problem where the metal tip of the electrode passed through the glass capsule to contact the muscle tissue. Schulman and his group had experience working with ceramics; the Clarion cochlear implant they had developed was housed in a ceramic casing. As a result, they decided to work on the leakage problem using a ceramic enclosure. This brought about a split in the three-way marriage between the groups headed by Loeb, Schulman, and Troyk, though the parting of the ways was reportedly amicable. Loeb and Schulman each began pursuing his own view of BION technology, in what Schulman called a friendly competition. Their familial ties remain, however. Both Loeb and Schulman are now affiliated with Alfred Mann organizations—Schulman with the foundation and Loeb with the institute.

In the final analysis, both teams were able to build hermetically sealed BIONs using their respective materials—Loeb glass, and Schulman ceramic. But the distinction between their BIONs does not end with the housing materials. Schulman's is a battery-powered system using lithium-ion batteries —each one hundred times smaller than a AA battery—that were developed by Quallion LLC, a company founded in 1998 by Mann. Though they require regular recharging by means of electromagnetic waves transmitted through the patient's skin, these batteries are expected to last many years before requiring replacement. Radio waves are also used to send operational commands to these BIONs.

Loeb, on the other hand, sends both power and stimulating commands to his BIONs via radio waves. Being able to send in enough power to stimulate nerves required some creative electronic engineering, since the amount of power that can be transmitted to the BION at any given moment is not enough to activate a nerve. The resolution was to send in bursts of power during the fraction-of-a-second intervals between BION stimulation of nerves. A unique capacitor design allows the BION to store energy it does not imme-diately use to build the power level. When stimulation is called for, the stored power is combined with that coming in at the moment to create an adequate current level.

Both the battery-powered and radio frequency–powered BIONs have their advantages and disadvantages. On the down side, the glass RF-powered device requires that an external coil be nearby at all times to power the de-vice. This can be ungainly when a large number of BIONs are needed to enable an individual who is paralyzed to stand and take a step, for example. The battery-powered BION does not require such a coil in close proximity, and operational commands can be sent to the implanted BION using a smaller transmitting device from farther away. Battery-powered BIONs can even be programmed to perform simple stimulation tasks with no external input. But the battery-powered devices are slightly larger than their RF-powered brethren, must be recharged every few days, and the batteries may eventually need replacement.

The innards of Loeb's glass BION look something like a wrap sand-wich made for an ant. The meat of the sandwich is an integrated circuit chip, which is surrounded by a ceramic covering. All of the ingredients are wrapped in a coil consisting of two hundred turns of wire about a quarter the thickness of a human hair, which serves as an antenna to receive the radio signals from outside the body. Protruding from the "wrap" are contacts at either end that connect to the electrodes passing through the glass enclosure into the muscle. The entire device, protruding electrodes and all, measures just over a half-inch long and less than a tenth of an inch in diameter.

In addition to creating all of the electronics involved, Loeb was con-fronted with the classic ship-in-a-bottle quandary: how to get it all into a hermetically sealed glass capsule? The solution was to leave holes in the ends big enough to get the ship inside the bottle and to place washers inside at both ends. Once the electronics are safely inside the capsule, the intense heat of a laser beam seals the ends of the capsule. But, as is so frequently the case in science, resolving one problem created a new problem of its own. Specifi-cally, the heat of the laser beam could lead to damage of the electronics inside the BION by causing excessive expansion and contraction. The solution was to mount the guts of the BION on a spring that takes up the motion and prevents damage. And just as ships in a bottle are reputedly made by old retired seafarers working laboriously by hand, though some of the BION's components are produced by machine, the entire device must also be assembled by hand.

The first generation of these devices, dubbed BION I, can be used only for stimulation. A more advanced BION has bidirectional capability that can record signals from sensors, like touch receptors, as well as stimulate muscles. One of the first sensors now being developed for incorporation into the BION is an accelerometer, designed to let an external controller know which way gravity is pulling on an implanted limb. In an able-bodied person, this is determined by the proprioceptive system working in conjunction with sensors that determine the length of each muscle. This information is processed in the brain, which fires off action potentials, enabling the body to compensate automatically for the pull of gravity. When there is a breakdown between the muscle sensors and the brain due to spinal cord injury, this system doesn't work. An accelerometer, which as its name implies, measures acceleration, will in effect close that gap by measuring the force of gravity on a limb and provide the information to a controller that does the processing the brain otherwise would. All of this will be enclosed in the BION to let it know when to fire a stimulus to keep the patient on an even keel. The BION accelerometer uses two semiconductor elements that change resistance under stress. Their electrical resistance varies depending upon how far they are from each other, which is determined by the acceleration of the BION capsule in which they are contained.

Taking advantage of the scalability of BION systems, researchers are starting human testing in small steps, planning to work their way up to increasingly more complex implants. In the early tests, BIONs have been implanted from foot to tongue with varying but generally positive results.

In one of the first human applications, BIONs were implanted to mediate foot drop, a condition experienced by stroke and spinal cord injury patients who retain some leg function but whose feet drop when walking because they do not have voluntary control of the muscles that lift the foot.

This leads to an awkward, slow, and sometimes stumbling gait. Before implanting BIONs in these patients, investigators had to determine when in the stepping process the BIONs should be activated and how to activate them. This was resolved by strapping an external tilt sensor to the patient's shin. When the critical angle is reached, the sensor transmits a signal to a single implanted BION to start stimulating. After the foot swings out, the angle sensor turns the BION off.

Another early experimental application uses the BION to exercise and strengthen weakened shoulder muscles to prevent and in some cases reverse shoulder subluxation, a condition that frequently follows stroke in which the muscles holding the shoulder in place relax and cause the shoulder to drop out of its socket. Two BIONs implanted in the shoulder muscles (one in the deltoid, the other in the supraspinatus) are activated when the patient places a coil on the shoulder in the vicinity of the BIONs, which stimulate the muscles following a preprogrammed regimen. Similarly, the basic BION is being tested in stroke patients who have lost the ability to open their hands. The implants stimulate the wrist and finger extensors to provide muscle tone and prevent permanent contracture. This repetitive exercise regimen can be turned on while the implantee is reading or watching television. One of the test subjects was so successful at developing the ability to use his hand that he became ineligible for a subsequent rehabilitation study. But at least one individual found that the BION caused so much discomfort that it was removed. Yet most test subjects do not feel the device's presence in their hands and find that the muscle contractions they derive seem natural. They are pleased that they can receive physical therapy without the help of a professional. BIONs have also been implanted in patients who suffer from severe arthritis of the knee.

A report on the results of the shoulder subluxation and knee test patients stated that "all subjects who answered the questionnaire would recommend this therapy to others with the same condition as themselves, and credited BION therapy for improvement in their joint function." The same report noted that in eight of ten shoulder subluxation patients "significant decrease was found comparing subluxation levels before and after six weeks of BION therapy." In the five implanted patients with knee osteoarthritis, "function improved over the twelve weeks of BION therapy," and "pain decreased significantly." As a result of these improvements, one subject cancelled planned knee replacement surgery.

BIONs have also been tested as a means of relieving sleep apnea. In this case, a single RF-powered BION is implanted directly in the tongue. The control coil is placed under the pillow, and when it is actuated, the tongue moves so it does not block the patient's airway. To determine proper placement, early tests were conducted on pigs, whose tongues are close to human size. The next step was to stimulate the proper tongue muscle overnight in humans with temporarily implanted wires to confirm that stimulating the area does not cause discomfort that would interfere with sleep.

An unexpected lesson was learned with the first person to be tested in this manner. When the wires were inserted into his tongue and the stimulating current was first turned on, he appeared quite startled even though the current wasn't high enough to elicit more than a slight twitch. Yet as he slept in the laboratory, the researchers gradually increased the current without waking him. When he awoke in the morning, they turned the current much higher than he would have tolerated the night before, and he didn't seem to notice. The ability to build tolerance to electrical stimulation was a valuable discovery, but the test subject did not experience enough bouts of apnea in the laboratory to properly test the configuration.

The next consideration was how to coordinate BION activation with the patient's breathing. Here, Loeb followed the KISS—or Keep It Simple, Stupid—precept. Earlier efforts at developing sleep apnea implants utilized relatively complex methods of coordination with the respiratory cycle that didn't pan out. Instead, Loeb decided to simply time the patient's breathing and have the BION turn on and off for the same period of time—say three seconds on and three off. The idea is to have the patient's breathing synchronize with the BION rather than the other way around. In early testing, the timing method seems to work well.

The one application for which the BION has been approved for sale in Europe and is in clinical trials in the United States is to overcome urinary urge incontinence. This particular form of incontinence tends to occur in women and causes the urge to void every few minutes, even though their bladders are not full. The BION treatment for this problem stimulates the pudental nerve, which has an inhibitory effect on the bladder. Most young children instinctively discover this pathway when they are too involved in playing to go to the bathroom and they massage their groin instead, which stimulates the nerve and reduces the urge to void. The BION does essentially the same thing in a more continuous, socially acceptable, and sophisticated way. The implanted BION continuously stimulates the pudental nerve until the user is ready to void, activates an off switch, and the urge returns.

In the longer term, the opportunities for the BION are boundless. Completely implanted systems with sensors and stimulators may someday provide a smooth and unobtrusive way for the paralyzed to move hands, arms, and eventually legs as well. These applications remain far in the future, for BIONs are still little more than implantable on/off devices, yet the eventual creation of such sophisticated systems is far from science fiction. It is a real possibility that is being investigated by real scientists, who deal only in science fact.

12

Pins and Needles
in the Brain

The entire field of neural prostheses involves placing manmade objects in the body to interface with the nervous system. And in the order of things, the brain is the crowning confluence of the nervous system. It is where the nervous system begins and ends. The brain originates signals to enable thought and movement, and it receives signals from every sensory organ in the body—eyes, ears, nose, tongue, and skin—and sorts them into some orderly fashion so that we can make sense out of those signals and react to them. When you see a painting, or hear music, or taste a fine meal, or smell a fragrant flower, it is not really the sensory organs that enable you to experience those pleasures. They are merely the messengers that pick up the packets of information and transmit them to your brain, which is where the experience is interpreted.

So in the final analysis, when the goal is to provide neural prosthetic relief for a condition in which the sensory or motor system is not operating properly, one must interface the prosthetic system one way or another with the brain. A rudimentary way of doing so is to have a paralyzed individual voluntarily move a portion of his or her body that retains movement to actuate a system implanted in another part of the body, as is the case with a hand grasp implant that is operated by a shrug of the shoulder. The thought process and the neuronal firings that move the shoulder originate in the brain. But shrugging the shoulder is awkward and provides minimal information as to what the user wants to do with the implanted hand device.

The ideal way to take signals from and send them to the brain is to interact directly with it, which for the most part, involves implanting electrodes either on or in the brain, and therein lies the conundrum. Since the brain is the seat of all thought and the center of our being, the mind, if you will, implanting it with foreign objects is the most socially questionable facet of neural prostheses. And because the brain is far and away the most complex and least understood organ in the human body, it is the most difficult to safely implant. Yet despite these caveats, scientists are working to create systems that would make the implantation of the human brain commonplace among the disabled, and even more controversially, if someday the risk/benefit ratio makes it feasible, to enhance the natural abilities of the able-bodied as well.

IN A LABORATORY AT BROWN UNIVERSITY in Providence, Rhode Island, a monkey sits in front of a computer monitor intently playing a video game. The laboratory is that of John Donoghue, the head of the department of neuroscience at Brown, and a leading brain-computer interface—or as he prefers to say, brain-machine interface—researcher. To get to Donoghue's laboratory, one must navigate a maze of halls and stairs before arriving at a small outer room cluttered with shelves of computers and monitors, some flickering with tables and graphs as the undisturbed monkey plays away in an adjacent closet-sized room. One of the monitors is hooked up to a closed-circuit television system that can be switched to show the monkey from various angles as it performs its experimental tasks. She is seated at what in laboratory parlance is known as a "primate chair," which looks something like a metal lawn chair. Her chin is cradled on a bracket, similar to the type one places one's chin on during an eye examination. The monkey's eyes have an almost human quality as they peer at the computer monitor in front of her. At her mouth is a rigid straw that runs from a juice bottle mounted on a wall just outside the room.

The monkey is totally engaged in playing a video game in which her job is to manipulate a mechanical arm in front of her—a computer mouse for monkeys—to move a red ball on the screen. The mechanical arm is a horizontal device, with a scissors joint halfway down its length, allowing it to be moved back and forth and side to side. The monkey's goal is to get the ball to strike a moving target in the form of a black square. Each time the ball successfully contacts the square, the monkey receives a reward of juice through the straw. It quickly becomes clear that the monkey is quite adept at striking the target and getting her reward.

While the monkey quite soberly works the ball toward the target, a speaker system in the outer room broadcasts the staccato, machine gun–like firing of the neurons in its motor cortex as they initiate its intent to move her arm. The sound is a cross between hail pounding on a tin roof and radio static. The spikes of the neurons are being recorded by a minuscule array of one hundred electrodes implanted in her motor cortex. The electrodes are wired to a pedestal that protrudes from the monkey's skull. A cable attached to the pedestal carries the signals to the bank of amplifiers and computers in the adjacent room.

Unbeknownst to the monkey, the mechanical arm she is manipulating is periodically disconnected from the computer, and the ball is activated solely by the firing of the neurons being recorded in the monkey's brain. The shift from the monkey's arm to her "thoughts" activating the ball is not quite seamless; the ball tracks somewhat differently when the mechanical connection is disrupted. The monkey seems briefly confused, but in short order she compensates for the altered movement of the tracking ball—much as one would compensate for the difference in the handling of a car should a tire go flat—and the red ball is again on target, although there is a 20 percent drop in accuracy. There is also a marked slowdown in the monkey's arm movement, as though the animal realizes she doesn't have to work as hard to get to the target and receive her reward.

While the monkey is at work, researchers are also monitoring the pattern in which her neurons are firing and comparing it to her arm movements so they can build the algorithms that will eventually enable them to translate brain activity directly into smooth, normal motion in a robot, or even a paralyzed human limb. Donoghue envisions a time, albeit in the far distant future, when this technology may enable spinal cord–injured persons implanted with brain-machine interfaces, and electrodes in their muscles, to move so smoothly the paralysis will be indiscernable.

That distant dream came a significant step closer to reality when on June 22, 2004, a brain-machine interface system similar to that used by the laboratory monkey was implanted in a human being. The recipient of the system was Matthew Nagle, a 24-year-old quadriplegic. The system he received was built by Cyberkinetics Neurotechnology Systems Inc., of Foxborough, Massachusetts, a publicly traded company founded in 2001 by Donoghue and some of his colleagues. Dubbed the BrainGate neural interface system, it utilizes the Utah electrode array developed by Richard Normann, who heads the Center for Neural Interfaces at the University of Utah. Normann founded Bionic Technologies LLC to advance the development of the Utah electrode array. His company was acquired by Cyberkinetics in 2002.

In 2004, Cyberkinetics received FDA approval, under what is known as an investigational device exemption, to experimentally implant the BrainGate system into five paralyzed individuals. Nagle was the first. His motor cortex was implanted with a one hundred–electrode Utah array, which is about the size of a baby aspirin, albeit a square one, and as thick as a contact lens. If placed on a penny, it would only cover the word "Liberty." The base of the array sits on the surface of Nagle's brain, while the electrodes protruding from it, arranged in a ten-by-ten grid, each 1 millimeter long and thinner than a human hair, penetrate into his motor cortex. A bundle of one hundred gold wires no bigger than a piece of linguine runs from the array to a thimble-sized pedestal that protrudes through Nagle's skull.

THREE YEARS PRIOR TO BEING IMPLANTED, Nagle, of Weymouth, Massachusetts, near Boston, was at a Fourth of July celebration at Wessagussett Beach in Weymouth, when an altercation broke out. Nagle, who weighed in at 6 feet, 2 inches, and 180 pounds, and who had been a football star at Weymouth High School, went to the aid of a friend. A knife was pulled and plunged into Nagle's neck just under his left ear. The knife entered his spinal cord at the C4 (cervical) level, instantly rendering him a quadriplegic, unable to move below the neck or even breathe on his own. His father, a Cambridge police detective was told that night that his son wasn't expected to live. But Matthew is a fighter. There were several surgeries to restore his ability to speak, and according to a report in *Wired* magazine, months of rehabilitation before he was able to move back to his parents' house. When he learned of the BrainGate human testing program from his mother, who ran across information about it on the Internet, he knew he wanted the implant, and he would not be deterred.

Nagle's implant surgery was conducted by Gerhard Friehs, a Brown University neurosurgeon, in a procedure that took about four hours. Magnetic resonance imaging was used to identify the precise part of Nagle's brain that would normally animate his left arm. The electrode array was placed in that region because the functions it was hoped he would be able to perform were those that would ordinarily require arm movement. A hole about the size of a half-dollar was drilled in Nagle's skull, then a specially designed pneumatic, high-velocity insertion tool was used to shoot the array into Nagle's brain in under 200 microseconds. "At that insertion speed, the brain basically doesn't respond to the fact that it has been implanted," said Normann, who invented the electrodes and the insertion tool. He compared the phenomenon to inserting a staple into a piece of wood. "One generally cannot push a staple into a piece of wood," he noted, "but if it is given sufficient velocity and momentum, the staple can easily be shot into the wood without damaging the wood or the staple." In the case of electrode insertion, since electrodes penetrate the cortex, it is impossible to prevent all tissue damage, yet an adult over 35 years of age naturally loses more neurons in a day than are killed by the insertion of an electrode array, said Donoghue.

After the array was shot into Nagle's brain, the gold wires from the array were attached to the connector pedestal secured to his skull. To use the system, a technician must attach a cable that runs from the external processing equipment, which fills a computer rack, to the connector. Each time Nagle wishes to use the system, it must be recalibrated by the technician, since micromovements of the electrodes can put them in contact with different neurons and because the firing patterns of individual neurons can vary from day to day.

Neurons relay information based on the rate at which they fire, which can fluctuate from ten to a thousand times per second but is usually in the range of hundreds of times per second. Yet a single neuron does not always fire at the same rate, even if the intent is seemingly the same. One day, a particular motor neuron involved in arm control may fire twenty times per second for a movement to the right, and the next day the spikes may be eighteen for the same movement. The reason for this variation is not well understood. Some experts believe it has no significance. Others think it does, that the cells may fire differently because of changes in motivation, or hunger, or that the learning process itself alters the firing rate. Whatever the reason, the early version of BrainGate requires that a technician recalibrate the system each time it is to be used to adjust for the daily variations. This, of course, does not leave the patient free to use the system whenever he or she chooses. Using Nagle as a "test bed," Cyberkinetics scientists were at work on a software system that would do the calibration on its own. This, coupled with hardware that can transmit recorded signals through the skin, eliminating the need to connect the computer to a skull-mounted pedestal, would allow a user to operate the system anytime without a technician present.

When Nagle was first plugged in and his system turned on, the Cyberkinetics scientists watched anxiously to see if the neurons in his motor cortex were still firing normally after years of paralysis. If they weren't, their years of work would have been for naught. The concern was that a motor cortex that had been deprived of sensory input for so long, because of a spinal cord injury, might show at least some deterioration. "It would not have been unreasonable to think you would not have gotten any useful data from the first case," said William Heetderks, who headed the NIH Neural Prosthesis Program before moving to the NIH's National Institute of Biomedical Imaging and Bioengineering. The researchers had done considerable animal testing prior to implanting the first human, but that was conducted on able-bodied monkeys whose brains had not been tied to paralyzed bodies for years. But the researchers' concern was unfounded. Spikes indicating the firing of Nagle's neurons immediately appeared on a computer monitor as the staticlike sound of their activity emanated from a speaker.

"These are really the first definitive results that show that one still has a high degree of control over those cells. I think it's very exciting," said Heetderks, who subsequently watched as Nagle used his thoughts to manipulate a computer cursor around a monitor. With this system, Nagle was able to control a television, turn lights on and off in his room at the New England Rehabilitation Center, and even retrieve e-mail messages using a simplified program. The next step would be for the patient to create his own messages using a typing program. *Wired* reported in its March 2005 issue that Nagle demonstrated his ability to "skillfully" play the computer game Pong, manipulating the game's paddle across a screen by his thoughts alone. He has even been able to use his thoughts to open and close a motorized prosthetic hand.

Monumental though these achievements are for a person who is completely paralyzed, they do not come easily, nor are they smooth. A lot more needs to be accomplished before more advanced actions can be taken, such as smoothly moving a limb or thinking a wheelchair into motion. In the meantime, Nagle, who signed on for the BrainGate study because he saw it as a first step to walking again, had the electrodes removed from his brain shortly after the end of the one-year test protocol. Typically, volunteers for implant experiments agree to participate for a specific period of time. They have the freedom to ask that the device be removed at any point, and if at the end of the specified test period, the organization that is conducting the test offers to extend the trial, the patient has the option of doing so. Shortly before the end of Nagle's twelve-month protocol, Cyberkinetics company president Timothy R. Surgenor said of Nagle's test, "I anticipate we will offer an extended protocol." But shortly thereafter, the company announced that he would have the system removed. There was no explanation at the time as to why it was being removed other than that the one-year protocol was ending.

While the first recipients of the BrainGate system are quadriplegics who retain the ability to speak, Cyberkinetics subsequently received FDA approval to test the BrainGate system on totally locked-in patients, such as those suffering from amyotrophic lateral sclerosis (Lou Gehrig's disease), who cannot communicate with the outside world. For them, being able to communicate by moving a cursor on a computer screen would make the risk/benefit ratio look even better than it does for high-level spinal cord injury patients who retain the ability to speak.

—⬩⬩⬩—

THE GENESIS OF THE BrainGate system goes back to when Donoghue spent five years in a wheelchair as a young boy growing up in Arlington, Massachusetts, not far from Boston. During the year it took to diagnose the rare bone disorder that disabled him, Donoghue spent a good deal of time examining X-rays along with perplexed radiologists. He vividly remembers one instance during that year when a physician studying an X-ray of his leg turned to the 7-year-old Donoghue, saying, "I don't see anything wrong here, do you?" At the time, Donoghue was unable to offer advice to his physician, but he became fascinated with the question. "Problem solving in the biomedical realm was something that really did catch my fancy," said Donoghue.

His malady was eventually diagnosed as Leggs Perthes disease, or aseptic necrosis of the femur, a degenerative disorder that affects the femur's head. The cause remains unknown, and the treatment is to keep weight off the leg to allow the bone to grow back, which it does, although not necessarily in a normal fashion. Donoghue was fortunate in that his femur did grow to be "more or less" normal. During his lengthy disability, Donoghue grew keenly aware of what it means to be disabled. "I went from bed to wheelchair to braces, so I got a feeling of what disability and medicine is about," he said. This sense of compassion was to stand him in good stead as his career evolved.

Following graduation from Boston University, with a bachelor's degree in biology, Donoghue went to work at the Eunice Kennedy Shriver Center, where developmental disabilities of the nervous system and mental retardation are studied. While there, he came under the wing of Verne Caviness, a neurologist who later became chief of pediatric neurology at Massachusetts General Hospital. Caviness became Donoghue's inspiration and mentor. Though Donoghue was only a technician, Caviness spoke with him and other technicians "as if we were colleagues at an equal level," said Donoghue. During one such conversation, when Caviness was discussing developmental patterns of the brain and abnormalities, Donoghue felt moved to tell him, "I don't have a clue what you're talking about." Rather than denigrating him for not knowing, Caviness provided encouragement by suggesting that Donoghue take a course on the subject being offered at Harvard, which he did.

"It was that kind of inspiration about understanding the brain that got me very excited," said Donoghue. It was also through Caviness that Donoghue met his future wife, whom Caviness was also mentoring. "He has completely dominated our lives, in the sense that he trained my wife and inspired me to go into neuroscience," said Donoghue. His wife went on to become a pediatric neurologist who works with children with spasticity caused by diseases such as cerebral palsy. It is quite conceivable that the kind of brain implant system John Donoghue has developed may someday be used as a treatment for some of her young patients.

Pursuing his growing fascination with the brain, Donoghue attended graduate school at the University of Vermont, where he studied neural plasticity—the idea that the brain can change in response to either damage or to learning. In particular, he examined structural reorganization in the brain following damage. While at Vermont, he toyed with the idea of going to medical school, but he had been bitten by the research bug. This led him to Brown University, where after completion of his master's at UVM, he focused on the organization of the motor cortex, the same part of Matthew Nagle's brain into which electrodes would be implanted a full twenty-five years after Donoghue received his Ph.D. in 1979. His dissertation research involved studying how the opossum's motor cortex is wired. It was during this research that he began learning how to record brain activity.

It is the cerebral cortex, or gray matter, that has expanded most during evolution and is sometimes referred to as the organ of civilization, since it is responsible for higher-order thinking. Donoghue was interested in how the cerebral cortex represents information and how it is modified. When he was a student, the accepted dogma was that the brain is fixed at a certain age. The idea that change constantly takes place in the brain was something of a revolutionary idea. Now it is accepted that the brain remodels itself all the time. "To me it is amazing to look back and think that anyone ever thought otherwise," said Donoghue. "How could it not be? Adults learn very well. There had to be some mechanism behind that." This new theory held the implication that the brain compensates for disease or damage, which encouraged researchers like Donoghue to believe that they could somehow promote the recovery of function.

It is a fortunate quirk of nature for those developing devices to implant in the brain that the cerebral cortex, which is made up of multiple layers of cells, has a modular organization. The fact that these modules are functionally localized—in that there are separate, physically identifiable areas responsible for each of our various senses and motor activity—is what makes it possible to implant electrodes in the brain that can record from or activate the cells involved in processing the specifically targeted function, such as motion or vision. Despite this modular makeup on the surface, most of the brain below the cortex is devoted to integrating all of the information it receives and generates. As Donoghue tells his students at Brown, "Eighty percent of the brain is involved in vision, the other 80 percent is involved in hearing, and the other 80 percent is involved in movement."

Following receipt of his Ph.D., Donoghue made the same career move that many leading neural prosthetic researchers have taken before and since. He went to work at the National Institutes of Health, the center of the neural prosthetic universe at the time, and a fertile field for innovation. Donoghue was thrown in with a group of scientists studying the physiology and anatomy of movement, and he found it highly stimulating. "Every hallway conversation was about the kinds of topics that interest me," he said. "It was absolutely wonderful."

The year was 1980, and Donoghue began two projects that eventually led to the invention of the BrainGate system. One was a continuation of his previous work aimed at understanding the processing of information in the cerebral cortex. He attempted to devise a method of using multiple electrodes to record from numerous neurons simultaneously to enhance this work, but despite four years of trying, he did not achieve this goal while at the NIH. His other experimental endeavor was to study the relationship between behavior and brain activity. He hoped that if one could understand how the brain formulates its intentions to move an arm, for example, then recordings could be made of the neurons firing as they signaled those intentions and the information could be used to operate prosthetic devices.

In 1984, Donoghue returned to Brown, this time as a faculty member, and continued his quest to record from the brain using multiple electrodes. He experimented with very fine wires inserted into the brain, but having a mass of wires protruding from a laboratory monkey's head proved ungainly and unsatisfactory. He felt there had to be a better way, and he found it in Richard Normann's one-hundred electrode Utah array. Donoghue met Normann at a professional meeting and was immediately impressed with his electrode concept. "The elegance of it was that it was a carefully manufactured device that consisted of one hundred microelectrodes in a little platform," said Donoghue. And though Normann's array was designed to inject visual signals into the brain, Donoghue felt it could just as readily be used in reverse, to record electrical activity from the brain. Normann was equally interested in Donoghue's work, the result being that the two scientists joined forces.

But there was still much work to do before Normann's electrode array could be safely implanted and record successfully for years on end. One concern, which proved unjustified, was whether the body would reject the array as a foreign object and push it up out of the brain. As it turns out, the electrode array does become encapsulated by scar tissue, but since the tissue forms on the top of the array, it helps rather than hinders the array's remaining in place. Another worry was that the brain's normal movement in the skull would cause the wires running from the array to the skull-mounted connector to pry the electrodes away from the brain. To limit the tethering forces, at first Donoghue only connected a few of the gold wires from the electrodes to the pedestal as he tested the system in monkeys. When the array held fast, he added a few more wires and continued doing so incrementally until all one hundred wires were connected without incident.

It took years to bring all the elements together, but by 2001, Donoghue felt he had all the parts for a neural prosthesis for humans. The data his team was getting from their monkey experiments enabled them to understand the intended movement of a hand through space. The surgical techniques they used to implant the monkeys were compatible with humans, and the electrode array demonstrated long-term safety and biocompatibility. The next step was to form Cyberkinetics to help obtain funding and governmental approval. Yet it still took three more years of preparation before it all came together and Nagle was implanted.

But Cyberkinetics was not the first organization to implant electrodes in the motor cortex of a human to record neuronal activity. That honor goes to Philip R. Kennedy, a clinical assistant professor of neurology at Emory University, in Atlanta. Kennedy invented a unique glass electrode that was first implanted in the brain of a woman suffering from advanced ALS in 1996. She died of the disease only seventy-seven days after surgery, but even in that brief period of time, Kennedy was able to capture clear signals from her motor cortex.

Like Donoghue, Kennedy founded a company—Neural Signals, Inc., of Atlanta—to further the development of the electrode, which is quite unlike any other. It's a hollow glass tube about the size of the tip of a ballpoint pen, containing two gold wires, which capture the firings of nearby neurons. Kennedy calls the device a neurotropic electrode, since it contains a neurotropic factor that encourages neuronal tissue to grow into the glass tube to help anchor the electrode in the brain. The wires run from the electrodes to an FM transmitter implanted under the scalp that sends signals to an external amplifier and computer processing equipment, with no through-the-skull connections required.

In March of 1998, Kennedy implanted two of his neurotropic electrodes in the motor cortex of a second patient named John Ray, a 54-year-old drywall contractor who was rendered unable to move and communicate by a brainstem stroke associated with a heart attack. Because Ray was in considerable pain and had numerous other medical problems resulting from his locked-in condition, he required frequent medication. This interfered with his attempts to use his thoughts to drive a cursor across a computer screen. But he did make progress. "I can't stress enough what a great job he's doing," Kennedy said at the time. "He could just say, 'The hell with you guys. I just want to lay here and listen to music,' but he knows that he's contributing and he wants to keep contributing."

Initially, Ray worked a program that allowed him to move a cursor horizontally and vertically. His task was to select an icon and hold the cursor on it for two seconds, whereupon it produced an audible response. Using this system, he eventually learned to spell out words at an agonizingly slow rate of about three letters per minute. To speed up the process, a group of Georgia Tech software engineering students developed a computer program that enabled Ray to point to complete phrases, such as "I'm too cold" or "I need the nurse." Four years after receiving his implant, Ray died of complications arising from his stroke, but throughout that period, the electrodes in his brain remained functional.

Kennedy then implanted a third patient, who suffered from mitochondrial myopathy, a disease of the muscle cells. Shortly after surgery, the disease attacked his brain, making it difficult for him to operate the brain-computer interface system.

Kennedy eventually decided to investigate less-invasive ways of recording brain activity than by implanting electrodes directly in the brain, and he set about creating two other types of electrodes. One, which he calls the Brain Communicator, is a conductive screw that is sunk into the skull deep enough to come close to, but not touch, the brain. This somewhat less-intrusive electrode records the firing of thousands of nearby neurons, much like the electrodes attached to the outside of the skull with an electroencephalogram (EEG)—a common medical device used to record brain waves—but, claimed Kennedy, his electrode is more precise. However, because the Brain Communicator is not as close to individual neurons as electrodes embedded in the brain are, the desired signals must be isolated from all the other firings going on, in a daunting process known as spike sorting. Yet some people who need a communication device but have the potential for recovery may well opt for the less-invasive approach.

The other type of electrode developed by Kennedy, known as the Muscle Communicator, is completely noninvasive and does not record brain activity. Instead, it is designed to capture the very weak electrical signals given off by muscle movement known as electromyographic signals, or EMG, which some locked-in patients can muster by moving a facial muscle slightly or twitching a toe. In some cases, Kennedy is combining signals from the Brain Communicator and Muscle Communicator to provide richer communication signals for locked-in patients. At the same time, he is continuing to refine the neurotropic electrode for use in conjunction with functional electrical stimulation systems to operate paralyzed limbs. Some experts feel the neurotropic electrode is too large to enable enough of them to be implanted to obtain sufficient signals for manipulating prostheses. Kennedy feels that a large number of electrodes will not be required and is working on shrinking the electrode.

There are several other major players developing brain-computer interface systems. At the University of Pittsburgh School of Medicine, professor of neurobiology Andrew Schwartz has succeeded in getting a monkey to feed itself using a robotic arm controlled directly by its brain waves. Like Donoghue and others currently involved in the field of neural prostheses, Schwartz took an early interest in the relationship between neuronal firing in the motor cortex and muscle movement. Following receipt of a Ph.D. in neurophysiology from the University of Minnesota in 1984, Schwartz obtained a postdoctoral fellowship in the laboratory of Apostolos Georgopoulos, a neurophysiologist at Johns Hopkins University, where he continued his studies on the motor cortex and arm movement.

Georgopoulos was investigating what was considered a revolutionary idea at the time, namely, that individual cells in the brain feed more than one muscle and that they fire for more than one direction of movement, in what is called a tuning function—meaning they fire at different speeds depending upon the intended direction of movement. These observations contributed significantly to the field of neural prostheses by suggesting that because individual cells perform many functions, it is possible to record from a relatively few of them and determine a person's intended movement. Georgopoulos even suggested that recording from between 100 and 150 random cells in the motor cortex would be sufficient to make a good prediction of where one was going to move in three-dimensional space. This work helped open the door to the development of implanted brain interfaces.

Not all experts agree that the motor cortex is the best place to capture neuronal firings that signal the intent to move. One renegade looking elsewhere is Richard A. Andersen, a professor of neuroscience at the California Institute of Technology, in Pasadena. Andersen argues that it would be better to record from the areas of the brain where visual feedback is processed and used to plan movement. These are the visual motor cortical areas, including the posterior parietal cortex and the dorsal premotor cortex located behind the motor cortex. It is in these areas that the eventual goal of movement is represented. If you want to take a jar of pickles off a supermarket shelf, for example, you first visually identify its location. The signals move from the visual cortex to the parietal cortex, where the plan to move your arm and hand is represented. The neural activity is then transmitted down the chain of command to the motor cortex, where the actual instructions to move your limb are given, and your hand obeys.

When Andersen began looking at using the visual planning centers for guiding prostheses, it was not known whether long-term paralysis would cause deterioration of the motor cortex. If it did, the parietal cortex, which depends upon visual rather than tactile input, would more likely remain intact. Since then, the implantation of Donoghue's patient Matthew Nagle seems to have negated the concern over motor cortex deterioration. Even so, since visual input is key to error correction, as in knowing when to change the direction of a cursor, Andersen still feels that recording from the parietal cortex may be preferable. And using signals from the planning areas of the brain before they reach the motor cortex has the benefit of making life somewhat easier for the software developers who translate the brain's activity into algorithms that activate prosthetic equipment. According to Andersen, if a thought is abstract, as in the planning stages, it can be interpreted to activate many different devices, but if it is more elaborate and detailed, you have to learn every representation for each device.

Dramatically, Andersen has been able to do a bit of mind reading while recording from the parietal cortices of monkeys, in what he calls "decoding cognitive variables." In a human, these are the signals that cause you to opt to receive $10 rather than $5 for performing a given task. In monkeys, the preferred reward is juice—as opposed to water—or a preferred kind of juice.

In Andersen's monkey experiments, an animal is trained to expect his preferred juice for successfully moving a cursor the first time. The monkey is also taught that the second time he successfully achieves the goal he will be given only water. The result is that the activity in the parietal cortex is much higher on the occasion that the monkey knows he will receive juice than when he knows he will get water.

Making the experiment more interesting, Andersen has trained monkeys to expect a given reward with a specific probability. Rather than every other time they perform a task, they may get their preferred reward a tenth, or a quarter, of the time. Then by reading the monkey's brain waves, he can determine what the monkey expects to receive before he receives it. Using brain waves to predict expectations could potentially offer caregivers of people who cannot communicate the ability to read brain waves as though they were body language to determine patients' emotional states or whether they are alert and paying attention. Taking the possibilities of this technology even further, it may become possible to read from the language areas of a patient's brain, which would do away with the need to manipulate a cursor to point to letters or words on a monitor. The ominous side of this "mind-reading" capability is obvious. Who of us would want our thoughts read?

The ability to read cognitive signals, and in fact the idea of recording from the parietal cortex, stems from Andersen's basic scientific quest for a better understanding of the brain. And, as he said, "the more we look, the more we find." The ability to read cognitive signals provides a case in point. "Being able to read out the brain signals and train the subjects to use them lets us learn how the brain learns new things. It also gives us a strong indication the brain is actually doing what we think it is doing."

Think about it!

13
From the
Inside Out

S cott Hamel is an automotive teaching assistant, who at 140 pounds held the bench press record for his weight class in New York State at one time by lifting almost twice his weight. Hamel also drives drag racers and owns his own drag race team. And he is a paraplegic, paralyzed in an automobile accident in 1977, when he was a junior in high school. But is he concerned about the danger involved in barreling down a drag strip at 200 miles per hour? "I never give it a thought," Hamel said. One thing he gives a lot of thought to, however, is operating a computer, which is something he does regularly, using his thoughts and nothing else.

On a bright, warm summer day, Hamel sits alone in his wheelchair in a darkened room in the massive Empire State Plaza in Albany, where the business of New York State is conducted and where the offices and laboratories of the State Health Department are located. He is wearing a bright red cloth hat, looking much like a shower cap, chin strap and all, except that it includes sixty-four small, white round dots symmetrically spaced across its surface, each representing an electrode that contacts the outside of his skull. Wires from the electrodes merge into a flat cable that leads to a computer on a table behind Hamel. Not moving a muscle, he is totally engrossed in the image on the monitor 5 feet in front of him as it responds to his thoughts, which are focused on moving the computer's cursor to a red oblong box that has just popped up in one corner of the screen. The other corners are populated by green boxes. It is Hamel's task to think the cursor to the red box, which requires moving the cursor up and down and side to side. When he successfully makes contact, the red box turns yellow and a new set of boxes appears. Where the red box will show up is arbitrary and differs each time the image changes. The faster he hits the target, the faster the screen configuration changes. If he does not reach the red target after a few seconds, the screen goes blank and a new set of targets appears. During the first of a series of three-minute sessions, Hamel hits twenty out of twenty-seven red targets. His score is similar for the other six sessions.

Unlike other systems, which utilize electrodes implanted in or on the brain to operate a computer with captured brain waves, the noninvasive brain-computer interface (BCI) system Hamel is testing records brain waves from outside his skull, using the same EEG technology that has been used as a diagnostic tool by the medical profession since the early 1900s. Electroencephalography was invented in the 1920s by the German psychiatrist Hans Berger, who used his 15-year-old son Klaus as a test subject. But it is highly unlikely that either father or son ever dreamed of the use Hamel is now making of the technology.

After demonstrating his dexterity, Hamel apologized for not doing as well as he had hoped. "On days I come in when I'm flying high, I will knock out thirty to thirty-four targets at over 90 percent," said the young-looking 44-year-old.

Trying too hard to think a cursor into motion is counterproductive, which is what prevented Hamel from doing his best during the demonstration. Because he wanted to do particularly well for a visitor, he did not feel as relaxed as usual. And once things didn't go as he had hoped, frustration set in, causing further tension. "Some days I come in here, I stayed up late, and I'm almost sitting in the chair taking a nap, and the cursor will fly all over the place wherever I want it to go," said Hamel. "I don't have to think about it. Other days, if I've had a tough day before I come here and my brain is very busy, or I get into a situation like I had today, it is a little bit of a struggle. It is the art of relaxing. On days I'm relaxed, and I don't give a damn if I hit the target or not, it just goes. It's like a basketball player when he can't miss a basket. He's just banging those baskets out. I find that zone; all of a sudden it's there. It flies."

The BCI system Hamel is testing was developed by Jonathan R. Wolpaw, chief of the Laboratory of Nervous System Disorders of the New York State Department of Health's Wadsworth Center. Wolpaw graduated from Case Western Reserve Medical School in 1970 and trained as a clinical neurologist before he decided to devote himself full-time to research. In what has become a virtual rite of passage for neural prosthetic researchers, he spent several years early in his career working at the NIH. It was there that he began his work on recording brain signals, first using the noninvasive EEG approach and then recording from the brains of monkeys implanted with electrodes. He also became skilled at using noninvasive EEG signals for therapeutic purposes, especially for biofeedback to help epileptics control the frequency of their seizures and for headache sufferers to control the intensity of their pain. From the NIH, and following a stint in the Army, Wolpaw moved to the New York State Department of Health, when in the mid-1980s, IBM and New York State began a joint project to develop new ways of using computers to assist people with disabilities. With his brain wave experience, Wolpaw was a natural choice to head an effort to develop an EEG-based brain-computer interface system.

His first challenge was to establish whether users of such a system could alter their brain waves rapidly enough to move a computer cursor in real time, since his previous work with patients using EEG involved producing gradual, long-term changes. A BCI system, however, would require instantaneous responses.

To establish the viability of the EEG concept, Wolpaw recorded from the scalp over the motor cortex—the same region in which invasive electrodes are implanted—where neurons fire to initiate movement. His goal was to use external electrodes to record mu and beta rhythms, two kinds of signals the brain gives off that are closely related to movement. These oscillations are the sum of the electrical fields coming from the activity of billions of neurons and synaptic connections in the cortex. They give off weak electrical signals of varying frequencies that can be detected by the electrodes attached to the outside of the scalp. Field potentials that run at a frequency of 8 to

12 hertz are called mu rhythms; those that fire at 18 to 26 hertz are called beta rhythms.

It took a number of years, but by 1994, Wolpaw and his colleague Dennis McFarland demonstrated that using nothing more than thoughts recorded by external electrodes, people could learn to control their brain waves rapidly enough to move a computer cursor in two dimensions—up and down and from side to side. But when he announced his findings, he was confronted by a nontechnical problem. Virtually no one was interested. That was because "there was no one around to notice," said Wolpaw. "There was hardly anybody else in the field." And that included both invasive and noninvasive methodologies. Though still in its infancy, the field of brain-computer interface has since made quantum leaps in growth.

Undaunted by the silence that greeted the announcement of their accomplishments, Wolpaw's group set about improving upon their results. Since the hardware they use is straightforward and not nearly as complex as electrodes that are implanted in the brain, they focused their efforts on upgrading the mu and beta rhythm signal-processing software. To make the system as user friendly as possible, Wolpaw's group created software that learns to follow the users as they become skilled at controlling their mu and beta rhythms rather than the other way around. The issue is "what we did wrong," not what the test subject did wrong, said Wolpaw.

To adjust for the fact that the frequency and location of the strongest signals coming from the brain differ from one person to another, the software adapts on the fly to different users. To do so, it continuously surveys the various electrodes that contact the skull to determine which are receiving the strongest signals, and it gives them the greatest influence in the instantaneous decision-making process as to which way the cursor should move. Accomplishing this required years of going through a complex iterative process in which investigators recorded a person's online activity and then tweaked the software offline. Complicating the process was the fact that what works offline doesn't necessarily work well when the user tries it online. Though Wolpaw's group has made significant advances, improving the software's sensitivity and response time and adding new features is a never-ending process. One of the more recent additions to the system's repertoire gives the user the ability not just to move a cursor to an icon but to click on it as well.

Hamel, one of several users helping Wolpaw perfect the EEG system, has been donning an electrode-laden cap at the Empire State Plaza laboratories three times a week for an hour at a time since 2000. Over the years, his ability to manipulate the cursor with his thoughts has improved dramatically because of his increasing skill and the ongoing improvements in the operating software. When Hamel first started with the project, he was limited to moving the cursor up and down. To move it up, he visualized curling the toes of his left foot. He found that by relaxing he could make it go down. "I would imagine my wife taking her fingernails and rubbing them on the palm of my hand. That would make me relax," he said. When he began trying to move the cursor from left to right, he imagined himself leaning from side to side, as though he was riding a bicycle and leaning into a turn, although he would remain erect in his wheelchair during the exercise. Eventually, the ability to move the cursor became second nature and he didn't have to "think" about it anymore. It came naturally, just as moving a mouse by hand comes naturally, without requiring contemplation. To "click" on a target, much as one clicks on a mouse with a finger, Hamel concentrates on squeezing his hand closed. This ability could eventually be used by quadriplegics to control a robotic arm.

Hamel volunteered to become a part of Wolpaw's team because of his interest in technology and his love of challenges. "The idea of monitoring my EEG to make something happen on a computer to me is pretty science fiction like," he said. "The control of my thoughts, the ability to make myself relax, I find that to be a pretty good challenge. When I walk out of here after a good day, there is a little feeling of satisfaction."

———

EVEN THOUGH THE USE OF EEG SIGNALS to operate a computer does not require surgery and is therefore safer than systems that utilize implantable electrodes, it has nonetheless become a controversial technology among many neural prosthetic experts. A number of them believe that it has a place in enabling totally locked-in patients to communicate by moving a computer cursor to point to icons or letters, albeit at an agonizingly slow pace, but the technology can never be precise and fast enough to drive a robotic arm, an implanted limb, or a wheelchair. This belief is based on the assumption that actuating such devices requires the ability to read specific neuronal firings, which cannot be accomplished by external electrodes because they also record the "noise" produced by the simultaneous activity of billions of other neurons going on at the same time. The selectivity just isn't there, they say.

But Wolpaw begs to differ. And his group has achieved some dramatic results to back him up. In fact, he and McFarland stood the field on its ear in late 2004, when they announced that they had achieved results comparable to those realized with electrodes embedded in the brain. A paper by the two researchers, published in the December 21, 2004, edition of the *Proceedings of the National Academy of Sciences* states, "We now show that a noninvasive BCI that uses scalp-recorded electroencephalographic activity and an adaptive algorithm can provide humans, including people with spinal cord injuries, with multidimensional point-to-point movement control [of a computer cursor] that falls within the range of that reported with invasive methods in monkeys. In movement time, precision, and accuracy, the results are comparable to those with invasive BCIs." As for whether comparing results achieved by monkeys and humans is valid, Wolpaw pointed out that monkeys are doing the same thing as humans, namely, trying to move a cursor to a target, and both get rewards. The monkey gets juice; the human gets to feel good. And "monkey's are very good at motor skills," he added.

Despite Wolpaw's accomplishments, which will undoubtedly open BCI to more applications, there is still skepticism in the field. William Heetderks, a former head of the NIH Neural Prosthesis Program who now directs the extramural science program at the National Institute of Biomedical Imaging and Bioengineering, is impressed with the amount of information Wolpaw has been able to extract from EEG signals and the rate at which he can do it. This, he feels, will open the door for EEG to be used for a growing array of clinical applications, including basic control of FES devices, such as those that open and close a hand implanted with electrodes. Yet Heetderks remains convinced that electrodes implanted in the brain provide richer recording than electrodes that record from outside the scalp: "Ultimately, if you want natural control that functions at the rate one would want to be able to pick things up and move them around, I think the only way to get that level of control is by cortical implants."

Wolpaw, on the other hand, maintains that results to date suggest that EEG-based BCI methods can provide such control. He compares the two methods to attempting to determine which way the wind is blowing by either examining individual molecules in the air or the atmosphere as a whole. "You don't necessarily have to go to the level of the individual molecules, and it may not be best to go to the level of the individual neurons to get control," he said.

Then there is the question of walking and chewing gum at the same time. Proponents of invasive electrodes claim that the noninvasive method of recording brain waves requires more intense concentration on the part of the user than do implanted systems. This would make it difficult if not impossible to operate a wheelchair, for example, or a robotic arm in the real world, where there are many distractions. "Even when an individual is doing a good job controlling the EEG, it really takes all of their mental effort to do so. They can't be distracted by anything. To really get a natural restoration, to pick something up off your desk without thinking about it, is going to have to involve implantation of electrodes into the brain," said Heetderks. He based this conclusion on studies done with monkeys "in terms of how readily they adapt the cells in their motor cortex to control computer screens or robotic arms. It's almost as if part of the brain is getting remapped to a new function. I don't think you'll ever get to that kind of result from EEG-based systems."

Wolpaw bristled at this suggestion, citing the fact that test monkeys with electrodes in their brains operate while strapped to a chair, which is a very constrained environment, and are likely doing a lot of concentrating. "There are no data to say that one requires more concentration than the other," he said. In fact, Wolpaw believes that neither electrodes implanted in the brain nor EEG signals have been proven to be superior to the other. He also questions the notion that signals taken from the motor cortex alone—regardless whether they are recorded by implanted or external electrodes—

can ever lead to what could be considered smooth, natural movement, which requires participation by multiple sites throughout the brain and spinal cord.

When neurons from one part of the brain, such as the motor cortex, are asked to do their job and the job of other neurons as well, they are being asked to change their natural behavior. Wolpaw equates it to asking the violin section of an orchestra to play an entire symphony without the other instruments. It will still be music, but not the same music. Likewise, operating an arm utilizing signals from the motor cortex alone is a different undertaking than doing so when the muscles are controlled normally, by the brain, spinal cord, and sensory feedback systems. The result is that when the cells of a cortical area are put in charge of the entire operation, the activity of the neurons themselves changes. "Hopefully they change so their performance in this new role gets better, but they change," said Wolpaw. Thus, operating such a system becomes a newly learned skill, just like playing an instrument. The expectation is that the more one does it, the more like a normal skill it becomes, but that remains to be shown.

In the end, however, the battle between invasive and noninvasive BCI systems may be a moot point, as a little-known compromise technology is coming along that may supplant them both. Called ECOG, for electrocorticography, this brain wave–recording system involves surgery, in that electrodes are placed on the surface of the brain, but it is considered less invasive than those systems that use pinlike electrodes to penetrate the brain. Though ECOG has been around for some time as a diagnostic tool, it has only recently been looked at for BCI applications.

ECOG electrodes are built into a flat sheet that neurosurgeons place on the brains of patients prior to undergoing surgery for epilepsy, for example, to determine where seizures are originating. Wolpaw and others have seized the opportunity presented during the few days or weeks that these electrodes are on the brains of surgery candidates to determine the quality of neural recordings they can achieve for BCI applications. The results have been encouraging. "It is clear that the signals we record there are similar to EEG signals, but much better," said Wolpaw. "The signals are much bigger, and they are not smeared by the skull and the skin." As a result, people with

ECOG electrodes on their brains can achieve cursor control in a fraction of the time it takes those using external electrodes. "What we've shown is that these signals have a lot of potential. In the long run, they may be the best solution for this kind of problem," said Wolpaw.

Wolpaw also thinks ECOG electrodes potentially present fewer long-term stability problems than other implanted electrodes because they are less invasive and therefore present less potential for scarring. ECOG electrodes can be placed on the pia mater, the thin layer of tissue that covers the brain, or even on the dura mater, the thicker outermost tissue covering the brain. If placed on the dura mater, the brain might not even know the electrodes are there and scarring would be minimal. And, noted Wolpaw, implanting such electrodes could be a relatively easy procedure. Sheets containing the electrodes could be placed under the scalp through a relatively small opening and then unfurled.

The reason ECOG has not gotten as much attention as other electrodes that record from the brain may have more to do with the sociology than with the technology of science. Many of those involved in the development of invasive electrode-based systems started out using implanted electrodes for basic brain research and then realized that the technology may have applicability in helping the disabled. Similarly, some of those working with noninvasive electrodes were originally involved in using EEG signals as diagnostic or therapeutic tools and later thought they might be able to help the disabled. In each case, they pursued what they knew best. ECOG got caught somewhere in the middle. But now that it has been "discovered," the technology is moving rapidly, and, according to Wolpaw, "there will be a good deal more heard about it in the near future."

In the meantime, Wolpaw feels that each of the various brain-computer interface approaches should continue to be developed, since each has advantages and disadvantages that have yet to be completely understood. He candidly points out that while his noninvasive method has the advantage of not requiring the surgical implantation of a foreign object in the brain, "contamination" of EEG signals by muscle or eye movement, can be a problem. And while the assumption that very little could be done with EEG signals has been put to rest by Wolpaw's work, how far they can be pushed to achieve greater BCI goals remains to be seen. Yet when it comes to the use of electrodes embedded in the brain, "we are still in the Stone Age," said Wolpaw. "We're basically putting beds of nails in the brain . . . It's a teeny bed of nails, but it is a bed of nails."

To help further the research in the field of BCI—both invasive and noninvasive—Wolpaw's New York State team, working in concert with experts from the University of Tübingen in Germany, have developed general-purpose brain-computer interface software dubbed BCI2000, which is being used by some sixty laboratories worldwide. By providing a basic software platform that all BCI researchers can use to set up their systems, receive signals, and analyze their results, BCI2000 prevents researchers from having to reinvent the wheel.

As the pace of BCI research and development continues to increase, there is little doubt that whether it be through electrodes implanted in the brain, or sitting on the brain, or on the scalp, or any combination thereof, wishing will, indeed, eventually make it move.

14

Reaching the
Depth of Depression

In a miraculous instant, a veterinarian who, because of Parkinson's disease, cannot stand on her own and whose hands shake so badly she has been unable to work is restored to almost normal movement. In that instant, the electrodes implanted deep within her brain are turned on, causing a charge of electricity to obviate her symptoms.

Another implanted Parkinson's patient finds that with his stimulator turned on, "the tremors aren't even in existence any more, and my balance is good enough to allow me to run. With the stimulation on, you can be close to normal. With it off, you can be close to vegetative."

In all, some 30,000 people have received deep brain stimulators to counter the effects of various motor diseases. The vast majority—about 90 percent—have been implanted to mitigate the ravages of Parkinson's disease, while the others became candidates for deep brain stimulation as a result of essential tremor or dystonia.

In addition to tremors, Parkinson's patients can have a problem initiating movements, called akinesia, in which they simply can't take a first step. They may also experience slowness of movement, or bradykinesia, and sometimes they simply lock up and are unable to move for extended periods of time. When drugs are no longer effective in reducing these symptoms, many patients are willing to accept the risks associated with having a foreign object implanted in their brain. And over 50 percent of those who receive deep brain stimulators find relief. Unfortunately, deep brain stimulation is not a cure for Parkinson's, and the disease continues to progress, but deep brain stimulation can give patients an additional five to ten years of relatively normal life.

———

THE MISFIRING OF NEURONS IN THE BRAIN that causes Parkinson's and other motor-related diseases is not well understood, nor is how electrodes in the brain alleviate the symptoms. What is known is that the subthalamic nucleus or the globus pallidus, two areas of the brain each about the size of a tic tac candy, are usually responsible for movement disorders. When electrodes implanted in these areas are turned on, an effect called neural modulation disrupts the pattern of abnormal electrical activity in the brain by influencing the behavior of whole patterns of cells, which brings the nervous system back into balance.

The salubrious effect of electrical stimulation of the brain was discovered serendipitously—as are many things in science—in the 1950s, when a surgeon was operating on the brain of a Parkinson's patient for an unrelated problem and accidentally damaged a blood vessel deep in the patient's thalamus. To his credit, he reported the mistake—and subsequently noticed an improvement in the patient's Parkinson's symptoms. Thus was the idea that making a lesion in the brain to mitigate Parkinson's symptoms born. The procedure's irreversibility was a drawback, however.

During the 1960s, physicians using electrical stimulation to locate sites in the brain for surgery not related to Parkinson's noticed that the symptoms of tremor in Parkinson's patients were diminished when certain areas were stimulated. Coupling the result of the accidental lesion with the observations of the incidental electrical stimulation led investigators to the idea of placing electrodes deep within the brain to essentially cause a temporary lesion by electrical stimulation. Two French professors, Alim-Louis Benabid and Pierre Pollak of the University of Grenoble, were the first to attempt deep brain stimulation for movement disorders, and they published the results of their work in 1987. They found that by stimulating at a high rate, symptoms were relieved, just as if a lesion was made, but the stimulator could be turned off, or even removed, if necessary. The electrodes also allowed for the adjustment of stimulation parameters to optimize effect.

The potential demand for deep brain stimulation was so great that it didn't take long by medical device standards for a system to become commercially available. The most widely used deep brain stimulation system is the Activa, produced by Medtronic, a large Minneapolis-based medical device manufacturer that received approval to sell the system in Canada, Europe, and Australia in 1995. Two years later, the Food and Drug Administration approved the Activa, which is similar to a cardiac pacemaker and has been referred to as a pacemaker for the brain.

The implant procedure itself can last up to eight hours, during which the patient must be awake so the neurosurgeon can determine by stimulation when the offending part of the brain has been located. The general area to be implanted is initially identified by means of high-definition MRI and CAT scans conducted prior to surgery. During the implant procedure, a clamp that screws onto the skull holds the head in place. This is apparently quite painful and has been referred to as a medieval torture device—although a system that uses infrared-sensitive cameras to align the head based on the MRI images is being developed. With the head secured, electrical leads are inserted into the brain through small holes in the skull. A computer program uses the MRI scans to guide the surgeon to the target area of the brain, but knowing precisely where to place the electrodes still requires a very human skill, namely, listening and interpreting subtly different sounds.

Neurosurgeons turn their trained ears to the spontaneous firing rate of neurons as they sink probes into the brain. As Andres Lozano, professor of neurosurgery at the University of Toronto and a leader in the field, said, "We are starting to understand the language of the brain and the language of neurons. It turns out that neurons in different places speak different languages, very much like people in different countries speak different languages. As we transverse [the brain] . . . we stop neuron by neuron and hear the languages of these neurons. By learning these languages, we have been able to pinpoint where we are in the brain and navigate through the complex territory of the brain to find the targets that are involved in producing the symptoms. This principle applies whether we are treating Parkinson's disease or depression, or anything else. Once we localize these targets, we are in a position to place an electrode in their midst."

In essence, said Lozano, "we are on a seek and control mission. Our role is to seek out the neurons that are misbehaving and to tell them to stop behaving in this way that is creating havoc. We are dealing with a brain dysfunction that is caused by a disorder in a few neurons that really plays havoc with and enslaves large cortical networks."

Electrodes are usually implanted in each hemisphere of the brain. Embedded wires are run from the electrodes up through the brain to just under the scalp. They are then routed behind the ear, down the side of the neck, and over the clavicle to a pacemaker-like stimulator, about the size of a cigarette lighter but as flat as a woman's compact case, that is implanted in the chest. The stimulator is then programmed to deliver a constant source of current to the target areas in the brain, thus modulating or disrupting the pathological behavior.

Deep brain stimulation has been so successful in treating the symptoms of Parkinson's and other motor diseases that it is undergoing testing as a treatment for epilepsy, chronic pain, and psychiatric disorders including depression and obsessive-compulsive disorder, and it is even being considered as a treatment for obesity.

Epilepsy affects over two million Americans, about a third of whom continue to experience seizures despite various treatments. When their seizures are severe enough to require surgery, the conventional treatment is to destroy the cells involved in the seizures, but in some cases even that has little effect. Implanting electrodes in the anterior nucleus of the thalamus, considered to be the crossroads where seizures originating in other parts of the brain converge and then propagate, may work where all other treatments have failed. And, as with Parkinson's disease, electrical stimulation has the advantage of eliminating the need to permanently destroy cells in the brain. Additionally, electrodes can be used to record from neurons as well as to stimulate them. This allows monitoring changes in the electrical activity of the brain that herald an oncoming seizure. Reading these signals and using the same electrodes to stimulate the brain to prevent the pending seizure may prove to be a simpler and more effective approach than constant stimulation.

A study of deep brain stimulation for epilepsy using the Medtronic Intercept Epilepsy Control System, similar to the Activa system used for Parkinson's disease, was begun in 2004. It includes 124 patients for whom at least three drugs have proven ineffective and who have experienced six or more seizures a month. Another brain stimulation system for epilepsy developed by Neuropace, called the Responsive Neurostimulator, is also undergoing clinical trials in eighty patients, which also began in 2004.

While the placement and firing of electrodes deep within the brain to counter epileptic seizures is a relatively new phenomenon, electrical stimulation of another part of the nervous system to reduce epileptic seizures was approved for clinical use by the FDA in 1997 and has since been implanted in over 31,000 patients worldwide. The VNS Therapy System developed by Cyberonics, of Houston, targets the vagus nerve as it runs through the neck. The vagus nerve is the largest in the autonomic nervous system, has a wide range of sensory and motor functions—including those associated with the stomach, pharynx, and the vocal cords—and connects to numerous parts of the brain. It runs on both sides of the neck from the various organs to the brain. Less invasive than a deep brain stimulator, the VNS (vagus nerve stimulator) electrode is implanted on the left side of the neck and delivers continuous pulses of electricity to the nerve. The controller, about the size of a stopwatch, is implanted in the left side of the chest, and a lead wire is run to the electrode in a one- or two-hour procedure. When turned on, it generates fifty electrical pulses per second for about one minute, stops for about two minutes, and then repeats the cycle.

Not a cure-all, the VNS device is used by many patients in conjunction with drugs to reduce seizures. Still, approximately 40 percent of those implanted with the VNS have experienced a 50 percent or greater reduction in seizures up to two to three years following implantation. And, according to Cyberonics, "seizure control and quality-of-life benefits with VNS therapy have been shown to increase over time. And over time, some patients have been able to reduce the dosage or number of their antiepileptic drugs."

The most controversial application of neural prosthetic implants involves the use of electrodes in the brain as a treatment for psychiatric illnesses, which conjures mind control and the nettlesome ethical considerations that go with it. Though experts disagree as to its effectiveness, vagus nerve stimulation is already being used as a treatment for depression in Europe, Canada, and the United States.

In an investigational study of the VNS for depression, 235 individuals received the implant. In half of the test subjects, the stimulator was turned on, and in the other half it was not. Three months later, the results were compared. Only seventeen of those patients in whom the implants were activated showed significant improvement, while eleven of the half who received no stimulation also reported significant improvement. There are two ways of looking at this evidence. One view calls into question the efficacy of the implant to treat depression, while the other takes the results to mean that the act of implanting the electrode itself may have some sort of beneficial effect. Despite the questionable results of the study, the FDA approved the use of the VNS for the treatment of chronic or recurrent depression in a controversial July 2005 decision. The device is also being studied as a potential treatment for other mental disorders, including anxiety, Alzheimer's, chronic headaches, and bulimia.

The more invasive treatment involving deep brain stimulation is also undergoing human testing for psychiatric disorders including depression, obsessive-compulsive disorder, Tourette syndrome, drug dependency, and schizophrenia. The potential for this therapy is dramatic, since it is estimated that depression is a primary cause of disability worldwide, affecting some 120 million people. And of the nearly 800,000 suicides annually, most are thought to result from depression. This is caused in large measure by the fact that 10 percent of depressed patients do not respond to any treatment, and many of them end up as suicide statistics.

Researchers are attempting to alleviate this suffering by targeting several areas of the brain for electrode placement when all other forms of treatment fail. Some are implanting the nucleus accumbens, which is associated with addiction and reward. Another group headed by Lozano, the Canadian neurosurgeon, and Helen Mayberg of Emory University in Atlanta, have targeted the subgenual cingulate region (known as area 25) in six severely depressed patients. They selected this region by conducting brain-imaging studies. First working with emotionally healthy individuals, they asked them to think of sad occurrences in their lives, such as the death of a parent. They found that the activity in area 25 increased in response to these thoughts, in what Lozano referred to as the "sadness center." They also found that when depressed patients were successfully treated with pharmaceuticals, the activity in area 25 decreased.

This led the researchers to ask two questions. They wondered if they could use deep brain stimulation to turn down the activity in area 25, and if so, would it alleviate depression? To answer these questions, they performed a ten-patient study. The patients they selected had been severely depressed for at least twelve consecutive months and had failed all medical treatments, including electroconvulsive therapy and at least four different drugs. "These patients we thought were the worst of the worst . . . unable to experience pleasure, and were suicidal in some cases," said Lozano.

That the researchers are on the right track became apparent even while the deep brain stimulator implant surgeries were in progress. The patients, who were awake during surgery so they could help guide the surgeons in their placement of the electrodes, reported phenomena like seeing the operating room fill with color or feeling a sudden sense of calmness when their brains were stimulated. As Lozano put it, "We have an acute effect that guides the placement, and based on that, we are able to know we are in the right place or whether we need to adjust the position of the electrodes."

Once the electrodes were implanted in the test patients and turned on, the operating room observations were verified by brain scans showing that the activity in area 25 was in fact significantly reduced. The investigators also noted that other areas of the frontal lobes that had not been active in these patients "came back online." As for whether the change in brain activity eliminated the depressions, one of the first six patients to be implanted spoke of relishing going out during the day for the first time since the depression set in, sleeping well, waking rested, and having a renewed interest in "everything and anything." That patient has since set up a successful business. In all, four of the six patients have had some response to the implant.

   Exactly how the electrical stimulation of area 25 disrupts pathological activity is not yet understood, though a predominant theory is that the stimulation causes the neurons to release certain necessary neurotransmitters. Perhaps once the mechanism by which the electrodes have worked for the four successful patients is understood, the neuroscientists will be able to induce positive results in the two patients who have not responded, and in the thousands upon thousands of other depressed patients who might be able to have their lives returned to them through deep brain stimulation.

15

A Hole in the
Center of the Brain

The eventual answer to Alzheimer's disease may be a chip in the brain. The same chip may also restore the lost capability to form and recover memories in those who have suffered brain damage from stroke or disease. At least that will be the case if Theodore W. Berger, director of the Center for Neural Engineering and a professor of biomedical engineering and neuroscience at the University of Southern California, has his way. Berger has been aiming one way or another at the creation of such a cognitive chip since his graduate days at Harvard, where he earned a Ph.D. in neurobiology in 1976. Putting his goal in real-world vernacular, Berger said its purpose is to reverse "neurological conditions that result from some unfortunate event that produces a hole in the center of the brain."

The primary focus of his work is the hippocampus, which is many synapses removed from the sensory input and motor output portions of the brain where other neural prosthetic researchers are working. Shaped like a banana and the size of a thumb, the hippocampi—each of us has two of them—are located deep within each hemisphere. They are involved in the formation of new memories and in the conversion of short-term to long-term memories. Berger has chosen to work first in the hippocampus before tackling the restoration of functions performed by other parts of the brain, such as those responsible for the processing of language, because the hippocampus is an evolutionarily primitive portion of the brain with a relatively simple cell structure. If one takes three slices of the hippocampus, the cells in each slice from top to bottom connect with each other. Other segments of the brain have a more complex cellular alignment structure, with cells connecting in a maze, much like a hodgepodge of wires plugged into an old-fashioned telephone switchboard.

The memory implant that Berger is devising uses the hippocampus's neat cellular alignment by acting as a bridge over damaged cells to reconnect the healthy cells on either side. It is Berger's hypothesis that it is not necessary to know specifically what the damaged cells did in the memory-forming process in order to replace them. Nor is it necessary to understand what the action potentials, or pulse codes, that had been created by the damaged neurons mean in order to create a chip that replaces them. All one must do is provide a processing bridge to let the healthy cells on either side of the damaged area talk to each other through an implanted chip. The chip's assigned task will be to do what neurons anywhere in the brain do, that is translate incoming pulse codes into different outgoing pulse codes and then pass them on to neighboring neurons.

Every time a neuron is activated by a pulse from a neighboring neuron, a variety of molecular changes begin to take place. The pulses fire off at anywhere from ten to thousands of times per second, and it is the difference in timing between the firings that essentially creates and carries information among neurons throughout the brain. The impact of an action potential on a receiving neuron generally lasts longer than the pulse itself. While the state of the neuron is changing in response to the first pulse it received, along comes another pulse from another nearby neuron, causing additional changes. This process continues until the neuron is altered to the point that it is ready to send off signals to its neighbors. This highly complex, and little understood, process going on virtually simultaneously in the billions of neurons that make up the brain gives us consciousness, memory, and the ability to contemplate our own being, which translates into mind.

If a part of the brain—in this case the hippocampus—is damaged, the result is the hole in the brain to which Berger referred. The cells providing output to the now empty space have nowhere to send their signals, while the cells on the other side of the gap receive nothing. Berger believes he can replace this hole with an implant by applying statistical analysis to the workings of the brain. He acknowledges, however, that developing a complete mathematical model of how the brain forms new memories in the hippocampus is likely a century away. But he feels such a model is not necessary to create a successful cognitive implant. Instead, "you can reduce the problem to the substitution of an input and output transformation and you don't have to think in terms of developing a complete understanding . . . of the cognitive function that you want to replace," he said.

But to do what Berger proposes requires an understanding of how the cells in the hippocampus transform inputs to outputs. This, according to Berger, is doable, since it is not necessary to figure out how an input-to-output transformation relates to a particular memory or how the memory is stored. It boils down to reading an input pattern and then knowing what the resulting output pattern should be. This can be analyzed because every part of the brain has its own peculiar input/output transformation pattern. Hence, said Berger, "I can figure out what the signal is, and I can duplicate what the output is supposed to be in a predictive manner."

But just learning the hippocampus's various input/output transformations through experimental observation, while observing signals from the hippocampus of an implanted monkey, for example, would take an army of researchers a lifetime to figure out, even without trying to understand what the transformations mean. So instead of direct observation in a living organism, Berger turned to an inanimate computer and the cold, hard facts of mathematics for help.

However, to arrive at a point where he felt he could use computers to figure out the workings of the brain, Berger first had to spend many years working with the real thing. As a graduate student at Harvard, he recorded the activity of individual neurons in alert animals to learn how information is transformed from one area of the brain to another. He then used that information to make inferences about the functioning of the memory process. After postdoctoral appointments at the University of California, Irvine, and the Salk Institute, Berger moved to the University of Pittsburgh, where he was affiliated with the departments of neuroscience and psychiatry and where he first applied computer modeling to the functioning of living brains.

That Berger should turn to computational science in search of answers about the functioning of biological organisms is not surprising when one considers that the Poughkeepsie, New York, native began his college career as a math major at Union College. He later switched to biology and finally to psychology. By the time he graduated from Union and headed for graduate school at Harvard, he knew he wanted to combine each of these fields to conduct brain research. He explained that he liked the idea of "treating the human as a kind of machine, the brain as a computing machine that determines the output of the system, which is the human . . . So being able to quantify how computing elements can determine what a human can do was my version of how to look at psychology."

Berger took courses in cognitive approaches to understanding the brain, like psychotherapy, but found them "unsatisfying." "Using such approaches," he said, "you observe a behavior in a particular environment and then you invent a story," by which he meant that behavior can be explained by any one of a number of different theories, none of which can be tested. "You just formulate a hypothesis, but there is no way to test the hypothesis rigorously." He finds the same holds true in the more empirical field of neuroscience. "If you are recording from a cell in a behaving animal, you can formulate a hypothesis about how the activity of that cell reflects some global hippocampal function that is determining the behavior of the organism, but how do you test that?"

Berger wanted to establish the brain's predictability by numerically quantifying the meaning of action potentials. So he set about learning how to quantify "the temporal patterns that neurons display in response to particular inputs and predict how these temporal patterns move to the next area, to the next area, to the next area to drive behavior." He saw this as being quite different from what others studying the brain were doing, which consisted of examining action potentials and saying, "I think this is what they mean."

But before he could reduce the activity of the hippocampus to raw num-

bers, Berger spent twenty years or so developing a more thorough understanding of the flesh-and-blood brain. As his work progressed and became
increasingly complex, the expertise of scientists from many fields was required, so Berger put together a team of researchers from a number of
universities and laboratories, including his Center for Neural Engineering.
The memory chip project now falls under the umbrella of USC's Center for
Biomimetic Microelectronic Systems headed by Mark Humayun, the retinal
implant developer.

By 1990, Berger felt his team was ready to attempt to attain the goal he
set many years earlier, that is, to mathematically quantify the activity in the
hippocampus. Even at that point, Berger did not have the design of a memory chip in mind. Instead his interest was oriented toward basic research. It
was not until he was five years into developing measurements that it occurred to him that the ability to predict how firing patterns are transformed
by neurons might be turned toward the more practical goal of developing a
neural prosthesis to replace damaged parts of the hippocampus and eventually even more complex structures within the brain.

Berger's team began the quantification process by programming a computer to generate a random series of simulated neuronal impulses. To cover
all the possible input patterns, Berger made sure that the frequency range of
firings generated by the computer ran from the lowest to the highest possible, even though most neurons only see a small slice of such firings. The
next step was to feed those computer-generated firing patterns into a real
neuron and record how the cell responded, in other words, what its output
pattern was. This monumental task was made somewhat easier by the fact
that cells in a given region of the brain react much the same to stimuli, so
that the computer-generated firings did not have to be fed into every single
cell. The hippocampus, for example, has several layers, each of which contains hundreds of thousands of neurons. While the cells in each layer react
differently to stimuli, those within the same layer tend to react very similarly.
Without such commonality throughout the brain and body, medical research
would not have gotten nearly as far as it has.

The cells' responses to the computer input in the layers of the hippocampus were then fed back into the computer, which related the input to the

output of the neurons and predicted how they would respond to any given pattern. "That's not a hope, that is something we can do now," said Berger, who claimed the accuracy of his predictions of neuron reactions to be in the 95 to 98 percent range.

Using this information to help patients, of course, requires that an appropriate electrode be implanted in the hippocampus. Unlike electrodes designed for use with other neural prosthetic devices that can successfully record from or stimulate many neurons in a general area, the cognitive implant must interface with specific individual neurons. In response to this need, Berger's team created a vertical, multisided ceramic chip that has adjustable metal pads on either side that can be set to contact specific neurons. The number of pads required is unknown and may depend on the area of the brain and the function being replaced.

The chip itself is dumb; all it does is transmit signals. The brain of the operation is a microstimulator implanted in the skull just above the brain that contains the computer-generated neuron models. Input signals recorded by the chip embedded deep in the hippocampus are sent through microwires to the microstimulator, which creates the appropriate output signals and sends them back down to the chip.

Berger first tested his system on slices of the hippocampus in the laboratory, which is possible because brain slices can be kept alive for about a day by feeding them oxygen and glucose. Since then, animal testing has commenced, first on rats, then monkeys. And in these tests the implant has "worked extremely well," said Berger. But as close as a monkey's brain may be to a human's, it is still not human. So, what if a monkey's hippocampus proves not to be a very good model for people? Berger is counting on two kinds of adaptability, one human, the other machine. On the human side, he is relying on the plasticity of the brain to adapt to the workings of the cognitive implant, as it does in taking the rudimentary input from a cochlear implant and converting it into recognizable sound. As for the machine, he expects the computer model used to mimic damaged neurons to adapt as well: "We can reprogram the chip from outside once the electrodes are in. If it turns out that the rest of the brain doesn't adapt enough, we can try out different parameters until the behavior of the patient improves."

Berger conservatively estimates that the memory chip will be ready for human testing by 2015 but hopes it will be sooner than that.

In an indication of the vast potential of this technology, the Office of Naval Research has asked Berger's group to create a speech-recognition system based on their computer modeling of the brain. The hope is that by simulating the action of the brain, such a system will be able to detect speech in noisy environments, just as humans can but existing speech-recognition programs have difficulty doing.

The hippocampus computer model promises to accomplish this because it looks at immediate past history—that is, the input signal—and rapidly transforms it to an output signal. Human ability to understand speech works much the same way; the brain breaks sounds down into segments. To recognize an entire word, all a person must do is hear a part of the word. Existing speech-recognition programs, on the other hand, examine all incoming sounds, so that extraneous noise can sometimes cause humorous mistakes, as anyone who uses such a program knows.

To create speech recognition based on temporal patterns of input and output, Berger's cognitive implant group records numerous people speaking the same word and teaches the computer to recognize all of the parts of the word as spoken by different people. It can then pick up on a part of the pattern of a word—the input—and use it to create the entire word—the output. According to Berger, during testing, the system "performs really well," even in environments where there is "lots of noise."

Surprisingly, said Berger, "our system has nothing to do with speech. It's just a model of how neurons change input to output patterns, which means you should be able to get it to recognize anything that has temporally coded information." Which means it has the potential to be used anywhere in the brain, for good or evil.

16
Ethics

Throughout history, there have been those who have voiced concerns about virtually every step in man's relentless quest to control his environment, cure diseases, and generally improve his lot. Yet, for better or worse, nothing has prevented the forward march of technology. The industrial revolution brought more goods to more people, along with sweat shops. Unleashing the awesome power of atomic fission brought relatively clean atomic energy, and the atomic bomb. Medical research has eradicated many diseases, cured millions of people of illness, improved the quality of life, and extended the lifespan of many in the industrialized world.

Yet in its quest to overcome the maladies plaguing humankind, medical science has also taken some wrong turns, as in the case of the Tuskegee Syphilis Study, revealed in 1972, in which the U.S. Public Health Service and the Tuskegee Institute studied nearly 400 black men infected with syphilis who were never told they had the disease nor were treated for it. And there were the infamous Willowbrook experiments conducted between 1963 and 1966 at the Willowbrook State School for children with mental retardation on Staten Island, in which residents were deliberately infected with the hepatitis virus to study an inoculating agent. Parents gave their consent, but they were told their children were being given vaccinations.

Though safeguards have since been put in place, the field of neural prosthetic research raises numerous ethical questions of its own. Yet in the case of neural prostheses, it is difficult, if not impossible, to find a researcher who has not to one extent or another considered the ethical implications of what he or she is doing. This is at least in part because of the fact that numerous groups are getting a jump start on the social implications of this vast new field by holding meetings, seminars, and panel discussions among scientists and bioethicists—whose numbers are rapidly increasing in hospitals, universities, and think tanks—to discuss the implications of the technology.

Questions being considered by these experts include: Should some facets of neural prosthetic technology, such as implanting electrodes in the brain, be pursued at all? When is an implant ready for human testing? How can researchers ensure that human test subjects are made fully aware of the potential risks involved? Once these devices are approved for clinical use, who should receive them? And who should pay for them? And, should they be used not only therapeutically to assist the disabled but also to enhance the natural abilities of the able-bodied, for example, to increase memory, instantly download knowledge—like a new language—or expand the wavelengths one can see and hear?

Mary Faith Marshall, professor and dean for professional conduct and humanities at the Center for Bioethics of the University of Minnesota Medical School, addressed these questions at a 2005 conference held at the Library of Congress entitled "Hard Science, Hard Choices: Facts, Ethics & Policies Guiding Brain Science Today." Marshall warned her audience of physicians, neuroscientists, bioengineers, and ethicists that even though they are attuned to the ethical issues surrounding neural prosthetic research "and take them very seriously, we can also go very seriously wrong." This can happen when well-meaning but overzealous researchers overlook the well-being of the individual in the quest for knowledge that could benefit many. "When you are applying a standard of care to an individual, you put the interest of the individual first. Within the research context, that doesn't always work when we are searching for new knowledge," she said.

Marshall advised that avoiding excessive exuberance does not mean one can or should eliminate emotion from decision making. "People who do that are psychopathic in some sense," she said. However, emotion should not overrule reason. This holds true both for researchers and human test subjects, who may be desperate for a cure and swayed to overlook risks by any suggestion of benefit. The question of how best to obtain truly informed consent from test subjects is a major concern among neural prosthetic investigators. Before this issue is dealt with, however, the decision must be made as to when a particular neural prosthetic system is ready for human testing.

While there are many hurdles, both governmental and institutional, that developers must overcome before obtaining approval for such testing, in the final analysis, it is the researchers' decision when to apply for approval and their job to make the case that a device is ready to be implanted in humans. In fact, the NIH policy specifies that the ethical responsibility for medical experiments lies with a study's principal investigators. Yet there are significant differences of opinion among neural prosthetic investigators over when a particular implant technology is ready for human testing. To some degree, the debate is divided into two camps, physicians in one and engineers in the other. Physicians, who by necessity frequently make rapid decisions about patient treatment and by nature are more action oriented, tend to come down on the side of earlier human testing. Engineers, experienced at carefully examining every potential problem, tend to be more cautious.

Philip Troyk, a professor of bioengineering at the Illinois Institute of Technology in Chicago, who heads a project aimed at developing a neural prosthetic system to implant in the visual cortex of blind individuals falls on the conservative side of the issue. When Terry Hambrecht retired from the NIH Neural Prosthesis Program, Troyk took up the visual prosthesis project that Hambrecht had been running internally. Hambrecht, a physician who also has bachelor's and master's degrees in electrical engineering from Purdue and MIT, respectively, had already experimentally implanted penetrating microelectrodes in the visual cortices of several humans. This followed years of safety testing in monkeys. Hambrecht and his team demonstrated that an individual who was totally blind could experience spots of light at precise locations in the visual field with electrical current levels a fraction of those previously reported by others with cortical surface electrodes. The last subject studied had only thirty-eight microelectrodes but could recognize simple patterns like letters when the appropriate microelectrodes were stimulated. Troyk, on the other hand, has not conducted any human testing since beginning his project in 2000 and does not plan to do so for some time.

"The project really moved from being focused on prosthesis development to looking at whether it makes sense to try to create such a system, and whether that system was technologically and functionally feasible for a human," said Troyk. He explained that he changed the direction of the project "because it is striking that despite almost one hundred years of knowing you can stimulate the visual cortex and get individual spots, no one has ever integrated those spots together into an image." So, he said, "it was not clear to anyone on our project that going to a human should be a goal per se. What we needed to do was step back and look at what the issues were from a variety of viewpoints, which were technological, physiological, neurosurgical, psychological, and ethical."

For one thing, Troyk is not convinced that the brain will be able to create coherent images out of spots of light. Other researchers feel otherwise. They cite the success of the cochlear implant, which allows the brain to create the perception of coherent sound out of relatively small bits of information. Troyk is skeptical that the visual system will do the same. "The visual system is sufficiently more complex than the auditory system, so that it is just not the slam dunk that the auditory system was from a functional standpoint," he said. "Vision is a perception that doesn't work by bit mapping. You don't have a TV monitor in your head." Troyk feels that before human testing is conducted, his team needs to understand more about how the brain breaks the signals received from the million nerve fibers in the optic nerve into their constituent parts and then processes them to create vision.

"We feel we are obligated to try, to the best we scientifically can, to understand that we can manipulate the visual system at the fundamental level. Then we will have something to offer the human volunteer. We can say, 'We don't know what the level of performance will be, but we are reasonably confident we know how to manipulate your visual system.' It's a much more sophisticated argument than just saying, 'Put it in, try it, and see what you get,'" said Troyk. But getting to that point, if it is ever achieved, can take a long time. Troyk's team is attempting to reach that level of understanding through animal testing. Hambrecht takes exception. He feels that Troyk's further experimentation in monkeys could "seriously delay the development of a visual prosthesis for blind humans. I feel that our NIH group's human experimentation answered essentially all the significant questions that might have been asked in monkeys and raised pattern recognition, stimulation interaction, and cognitive adaptation questions that can only be answered with more sophisticated implants in blind humans," he said.

Hambrecht also raises the question of how one can really know what an animal is seeing. He acknowledges that there are sophisticated methods of determining an animal's interpretation of simple stimuli based on rewards for specific behavior, but no one will really know what a blind animal sees when a complex visual image is presented to its visual cortex. Only a human can interpret and communicate this. "One of my other concerns about further experiments with monkeys is that researchers will misinterpret what the animals are experiencing and design a human prosthesis based on these misinterpretations," said Hambrecht.

Robert Greenberg, a physician and engineer who is the president of Second Sight, a company that has experimentally implanted chips in the eyes of a number of individuals who are blind (see chapter 10), agrees with Hambrecht. As Greenberg sees it, the primary goal of animal testing is to ensure the safety of an implant. Once that has been established, implants should be tested on humans. "There are things you can learn from animal studies, but at the end of the day you really can't ask them what they see. You can train them to press a lever if they see a flash of light, but that's a far cry from asking, 'What is the character of that light? Did you like that? Was that a pleasant experience or a bad experience? Is it useful?' "

Having said that, Greenberg thinks Troyk may be correct when the question of readiness for human implantation pertains specifically to the visual cortex implant. "We felt that for a retinal stimulator, we were implanting a blind eye with a device that was heavily tested, so the risk to the patient is quite low. They are already blind, so there was no risk to the vision, and the device was well tested." Implanting the visual cortex of the brain, however, is another matter.

When contemplating whether to implant, the question also arises as to what the goals of a prosthetic device should be. Most agree it must compete with and even exceed existing assistive devices. In the case of visual implants, realistic goals would be to provide people who are blind with greater mobility and the ability to read. This means the technology is competing against canes, guide dogs, Braille, and computerized reading programs. As Troyk put it, "If we can't do better than a cane or a guide dog, why ask someone to have a thousand electrodes stuck in their head?"

Troyk claimed that "what has been lacking is that no one has asked individuals with blindness, what would they like to have? What function do you value enough to have complicated and risky brain surgery done? That's an important point that I think the entire field of visual prostheses is lacking in, but it is a prominent point of our project. We are exploring this through components of our project in psychology and ethics. We have dedicated a fair amount of our efforts to those topics. The point being that today mobility and reading may not be compelling motivations for a visual prosthesis."

Yet most human neural prosthetic test patients, including those who have had implants permanently placed on their retinas or on their auditory brainstems, say that having some sensation, regardless how little, is far better than none at all.

———

THE FIRST OF MAN'S ATTEMPTS to protect human medical research subjects dates back to the Hippocratic Oath, which states that "I will use my power to help the sick to the best of my ability and judgment; I will abstain from harming or wrongdoing any man by it." Unfortunately, the oath has not always been followed. During World War II, Nazi researchers, including highly regarded physicians, conducted unspeakable experiments on men, women, and children in concentration camps. As a result of the 1946 Nazi Doctors' Trial in Nuremberg, an international code of ethics known as the Nuremberg Code was established the following year. It requires informed consent from subjects of medical experiments, that human trials be preceded by animal studies, and that the benefits to science be weighed against the risks to human subjects.

But the code's informed consent provision has not always been adhered to. In the late 1950s, it was discovered that a number of deformed babies, primarily in Europe and Canada, were being delivered by women who took thalidomide, not knowing it was an experimental drug. As a result, the 1962 Kefauver-Harris amendments to the 1938 Food, Drug, and Cosmetic Act put teeth in the code by making informed consent the law in the United States. Shortly thereafter, the FDA incorporated that requirement into its regulations. Four years later, the U.S. surgeon general issued a policy calling for institutional review boards (IRBs), asking all institutions that sponsor medical tests on humans to form a panel of experts to review and approve studies before they begin and to follow them as they progress. The FDA later established formal procedures requiring that research projects also be reviewed.

In 1972, the Tuskegee Syphilis Study, which had been going on for forty years, came to light. During the same period, the public learned of the Willowbrook Hepatitis Study. These miscarriages of medical research ethics caused a public outcry that resulted in the National Research Act of 1974, which created a commission of distinguished physicians, lawyers, and ethicists charged with developing ethical guidelines for the conduct of human research. The National Commission for the Protection of Human Subjects of Biomedical and Behavioral Research convened for four days of intensive discussions in the Smithsonian Institution's Belmont Conference Center. These discussions resulted in a report setting forth basic ethical principles for medical research involving humans. Now known as the Belmont Report, it has served as a touchstone for those involved in medical testing of humans ever since.

At the 2005 "Hard Science, Hard Choices" conference, over twenty-five years after the Belmont Report was issued, William Heetderks, a former director of the NIH Neural Prosthesis Program, cited the report's three ethical principles—beneficence, justice, and respect for persons—as "great guidelines" for neural prosthetic researchers.

Under the heading of beneficence, the report calls upon researchers to "(1) do not harm and (2) maximize possible benefits and minimize possible harms," while also stipulating that human testing should be "justified on the basis of a favorable risk/benefit assessment." Heetderks pointed out that when it comes to neural prosthetic testing, it is difficult to adhere to the "do not harm" dictum because, "we are talking about doing things that have the potential to do significant harm." On the other hand, he noted, maximizing the benefits and minimizing the dangers "is a realm in which many of these procedures can be done."

The justice segment of the report grapples with the question of "who ought to receive the benefits [of research] and bear its burdens." The report notes that in the past, "the burdens of serving as research subjects fell largely upon poor ward patients, while the benefits of improved medical care flowed primarily to private patients." To prevent such injustices from continuing, investigators are advised that they "should not offer potentially beneficial research only to some patients who are in their favor or select only 'undesirable' persons for risky research."

The report also calls upon those selecting test subjects to make distinctions between "classes of subjects that ought, and ought not, to participate in any particular kind of research, based on the ability of members of that class to bear burdens and on the appropriateness of placing further burdens on already burdened persons. Thus, it can be considered a matter of social justice that there is an order of preference in the selection of classes of subjects (e.g., adults before children) and that some classes of potential subjects (e.g., the institutionalized mentally infirm or prisoners) may be involved as research subjects, if at all, only on certain conditions."

The "respect for persons" part of the Belmont Report has great relevance to neural prosthetic researchers, since it speaks directly to the issue of informed consent—the strongest protection for neural prosthetic test subjects. The report's primary point in this regard is that "individuals should be treated as autonomous agents" who are capable of deliberating and then making their own decisions. But it qualifies that statement by making it clear that "persons with diminished autonomy are entitled to protection." This group includes people who are ill, have mental disabilities, or are prisoners.

Despite these guidelines and the FDA requirement that volunteers for medical experimentation sign a document stating that they have been informed of the risks of the test they will be subjected to and give their consent to be a part of the study based on that information, informed consent remains the subject of substantial discussion among researchers and bioethicists. What constitutes fully informed consent can vary significantly from institution to institution. And can patients with incurable maladies really make objectively altruistic decisions, even when told they will be putting themselves at risk with little hope of helping themselves, or does desperation override their intellectual decision-making capacity? As Illinois Institute of Technology philosophy professor Michael Davis put it, "There is a question of information and a question of emotions, and you can understand and still get carried away by your emotions."

In a novel approach to patient protection, Davis is one of two ethicists who serve on a panel of experts that meets periodically to discuss the progress of IIT's visual prosthetic project headed by Philip Troyk. Working with these ethicists, the group has decided that the best way to ensure informed consent is to have implant volunteers participate in the research, or at least sit in on the group's meetings, for a year or more before submitting to implantation.

"Part of what you want to do is scare prospective test subjects, get them to see how dangerous it is," said Davis. "They are essentially putting their lives on the line, and you want them to appreciate how risky that is, and that takes time. But also, it is partly a process of getting them to see how interesting and important the work is."

—⊸∰∯∰⊷—

LOCATED IN BUCOLIC GARRISON, NEW YORK, hard by the Hudson River, and 50 miles north of New York City is the Hastings Center, where philosophers and other academics spend most of their time contemplating the ethics of medicine. Situated in a majestic old Victorian house at the end of a dirt road, the center is high on a hill directly across the Hudson from West Point, overlooking Constitution Island, where George Washington's troops strung a steel chain across the river to prevent British ships from working their way upriver during the Revolutionary War. The magnificent surroundings foster the kind of thinking that makes the Hastings Center one of the world's foremost bioethics think tanks. The atmosphere inside is informal and collegial. Guests are asked to join staff members for colloquial brown bag lunches, at which they sit around tables set in a square to discuss their work, in something of an Algonquin Round Table for bioethicists.

On a sunny summer afternoon, the subject of discussion turned to the funding of neural prosthetic research, which comes in large measure from governmental organizations, including the NIH, the Veterans Administration, the National Science Foundation, and the Defense Advanced Research Projects Administration, an arm of the U.S. Department of Defense. Because these funds come from highly diverse and bureaucratic organizations, it is nearly impossible to tally the total investment. Even the NIH does not have an aggregate figure available for the amount its various programs spend on neural prostheses. Yet two points are clear. First, as the technology looks increasingly attractive for widespread clinical applications, the money made available to researchers is increasing. Second, the amount invested in neural prosthetic research and development through government, foundations, and private investment pales in comparison to the money poured into other medical research for pharmaceuticals and dramatic life-saving devices, such as a totally implantable heart.

This led Mary Ann Baily, a Hastings Center associate for ethics and health policy, to decry the fact that more has not been done to fund the development of neural prostheses. Baily, an economist, pointed out that the U.S. health care system seems more willing to invest in more spectacular life-saving technologies than in technologies that can help disabled individuals live higher-quality, more-productive lives. "We have a bias within our decision-making system around keeping people alive. If you can keep somebody alive on life support, you practically have to have an act of Congress to let you turn it off," said Baily. She then added, "To my mind society is going to get a lot more value for the money out of neural prostheses that enable people to function like normal human beings. People who are not in danger of dying in the near future but who simply have trouble functioning. We should be spending more money on them."

One reason for this disparity, noted Baily, is that as a rule, people with disabilities do not have economic clout. "There is a connection between being profoundly disabled in this country and not being very well-off, for obvious reasons. So there isn't that automatic market," she said. Another factor she sees at play is that competing insurance companies and government entities do not expect to cover any given individual for a long period of time because of the rapidity with which Americans change jobs. As a result, they are not amenable to investing up front in potentially costly technologies that will save money in the long run by reducing the need for assistive care. To overcome this tendency, the nation needs "a rational health care system, one in which there is some attention paid to the whole picture," said Baily.

Without higher levels of government funding or greater commercial interest on the part of established health care companies for neural prosthetic development, those committed to advancing the technology have little choice but to start their own commercial entities to attract venture capital— an important step toward bringing a useful prosthesis to those who need it. As Richard Normann pointed out, "Until a device is commercialized, until you've actually identified a company that can take the technology from what we've done to commercial products, you really haven't helped society at all." Normann formed Bionic Technologies—subsequently acquired by Cyberkinetics—to produce electrodes he developed.

For his part, John Donoghue, of Cyberkinetics, is enthusiastic about the corporate structure as a way of helping nurture neural prosthetic technology. "There's a different kind of discipline that occurs in a company than occurs in a laboratory, and I think a company has the right structure to take what has been found in science and move it forward," he said. "The idea of a company is extraordinarily valuable and essential. It's another source of funding, and the scale of funding is better. You can raise a lot more money."

But going commercial raises important ethical questions. Will the possibility of loss or profit encourage the overexuberance that Marshall warned against and cloud one's objective view of a technology? Arthur Caplan, professor of bioethics and director of the Center of Bioethics at the University of Pennsylvania, thinks that could well be the case. He suggests a dose of skepticism when considering the claims of commercial companies involved in experimental neural prosthetic development. "Just as was true in gene therapy, they have to push hard to attract funders," said Caplan. "These companies are so desperate to get out there running with a product—they are going to go out of business if they don't—they see benefit around every corner." This leads to "an observer effect. It starts to leak in and it's hard to avoid. It's going to create, if not a conflict of interest financially, a conflict of observation." Thus, "it would be desirable to have people outside the company both doing the animal studies and assessing and evaluating the human studies." Putting it into a nutshell, Caplan said, "If you own it, you shouldn't be studying it."

Caplan acknowledged that neural prosthetic scientists generally have little choice but to go to the marketplace to obtain funding for their inventions. "That is the process. You either attract capital that way or you've got to pull it out of foundations or patient advocacy groups," he said. And, he noted, academia is not free of similar temptations, except in that world it can be the competitive desire to be first and to publish that may color a researcher's view, rather than the lure of funding. To prevent against any such abuses, Caplan feels that peer review is extremely important.

"You have to have all your data transparent, you have to have anybody assess it who wishes to do so, you have to basically be open about presenting everything so that others can see what's going on," he said. "And in doing clinical assessments you definitely need independent evaluators and double-blind controls to see what's going on. These are methods that are well understood, there is nothing new about them, but they often get distorted, or usurped when the pressure comes on."

As for whether institutional review boards and FDA requirements are enough to protect human test subjects? "It is beyond dispute that they are not," replied Caplan. "They are overwhelmed, overworked, and they often have their own conflicts. The IRBs want institutions to do well in terms of research. You are often evaluating your colleagues. And the FDA doesn't have the tools to really do the job . . . I'm not arguing they should go away. I would like to see them beefed up in various ways. But from the point of view of implants, the more independent experimentation, the better, by people who don't have a direct fiscal stake in the outcome."

Illinois Institute of Technology philosopher Michael Davis has a similar view of FDA and IRB safeguards for neural prosthetic research: "They provide sufficient protection for normal experimentation, but when you are dealing with new technologies, groundbreaking experiments, they are going to be behind. That's just the way it is." The reason, he said, is that existing safeguards were not designed with breakthrough technologies in mind. "From the perspective of engineering ethics, this is the normal state of things. Governments and regulations are always behind the technology. That is one of the reasons it's important for engineers to exercise their own judgment on questions of safety within what is legally required."

Sharona Hoffman is an associate professor of law and of bioethics at Case Western Reserve University. She holds a Master of Laws degree in health law and has served on institutional review boards in Texas and Ohio, including the IRB for Cleveland's MetroHealth Medical Center, one of the institutions that is part of the Cleveland FES Center.

According to Hoffman, the IRB process for neural prosthetic implants is no different than for other proposals the IRB has to consider. "We evaluate the risks, we evaluate the potential benefits, and we make a judgment as to whether or not the benefits outweigh the risks and the protocol should be approved," she said. In fact, Hoffman has found the neural prosthetic proposals to be less ethically problematic than those involving very high-risk chemotherapy and AIDS treatments, where life is in the balance. "Not that many patients are at risk of loss of life or significant further disability from these devices," said Hoffman. Thus, "I don't recollect them as being more controversial than other types of protocols."

While there are guidelines for the makeup of IRBs, such as the requirement that they include at least one scientist, one nonscientist, and one person who is not affiliated with the institution, the number of members varies depending upon the size of the institution. The MetroHealth Center IRB on which Hoffman has served has fifteen to twenty members.

When a request for clinical trial approval is presented to the IRB, a member who is a specialist in the field is designated to oversee the review. He or she considers the proposal in detail and requests additional material from the researcher if necessary. All the materials, including the informed consent form, are then submitted to all of the members of the board for their review. A general meeting of the board is then held to discuss the protocol, following which a vote is taken, with a majority required for approval of the clinical trial.

Though IRBs examine the science to make sure it's sound and the risks are reasonable, "the IRB's function is not to determine the scientific validity of a trial, but rather the ethics of it," said Hoffman. The IRB must also do its best to make sure the institution does not get sued as a result of the study, which is one of the reasons its members pay careful attention to the informed consent process.

"If a person is hurt because of the study, there is a risk the person will sue the doctors, the hospital, the institution, even the IRB," said Hoffman. "Therefore we pay particular attention to the informed consent, and we make sure that every risk is disclosed, and that they make statements that there may be additional risks that haven't been revealed yet. That's why this is an experimental study. And we try to make sure that people are consented in a way that they understand what is being asked of them. If they consent, it's much more likely the institution will be able to establish that they waived liability."

The vast majority of clinical trial requests are approved, although on occasion the IRB defers its decision to obtain additional information or asks that the proposal be rewritten. The IRB's job is not done once a clinical trial is approved, however. It is also responsible for periodically monitoring the trial to ensure that all is going well. If a participant sustains injury, or if there is some other adverse event, the investigator is required to immediately report it to the IRB, which must consider if the risk requires altering the informed consent process or even halting the trial.

Some in the field are more concerned with what may happen when neural prostheses enter clinical practice than they are with abuses that may occur during the research phase. In fact, Robert Goodman, a professor of clinical neurological surgery at the Columbia University College of Physicians and Surgeons, thinks research should be easier to perform and clinical practice should be more restrictive. His specific concern is the surgical version of what is referred to in the pharmaceutical field as "off label use," or the use of a drug to treat conditions other than what the FDA approved it for. Speaking at the "Hard Science, Hard Choices" conference, Goodman pointed out that obtaining approval for a medical device requires controlled research and a major monetary investment "to prove that a surgical treatment involving an implant is both safe and efficacious." But, he noted, once it is approved, the same oversight does not apply: "Once it overcomes those hurdles and gets approved, like a drug, the approved procedure can be applied not necessarily under the same circumstances it was applied under to get approval. That is where we can run into trouble."

Goodman used vagus nerve stimulation to treat epilepsy—in which electrodes are placed on the vagus nerve in the neck—as a case in point. The vagus nerve stimulator is intended for patients who either are not candidates for surgical treatment of the brain, in which the part of the brain causing seizures is destroyed, or who do not want to undergo brain surgery. Yet Goodman, an expert in the field, finds that "in many cases, patients are being offered an implant with this device without either of those two indications." This happens when, because of overzealous marketing, the technology "is no longer applied through epilepsy centers, where physicians with experience using a surgical treatment for epilepsy are considering it for implantation," he said. Instead, vagus nerve stimulation "is actually promoted to be used by physicians in community practice as an adjunct to treating patients with epilepsy, and that includes neurologists without experience with surgical treatment."

To guard against implant abuses, Goodman recommended a physician team approach, "because when you're obligated to work in a team, and we have a conference to discuss whether a patient is the appropriate candidate for certain kinds of intervention, you have to convince your colleagues that you've followed the appropriate criteria."

The longest-running ethical controversy over the implantation of neural prostheses revolves around cochlear implants, the first neural prosthetics to be approved by the FDA. These devices, designed to return hearing to the deaf, were first approved for implantation in profoundly deaf patients in 1984. As experience established the efficacy and safety of these devices and the technology improved, cochlear implants were subsequently cleared for implantation in people who retained some degree of hearing and eventually for use in deaf children as young as 6 months. Follow-up studies of children who have been implanted at a very young age show that, for the most part, they tend to learn language at the same rate as hearing children. And because the sounds produced by cochlear implants are the only sounds that deaf children know, they tend to do better with their implants than older people who learned language with normal hearing prior to going deaf.

Yet there are those in the deaf community who have objected to cochlear implants on the grounds that they are a threat to deaf culture. During the early 1990s, the National Association for the Deaf (NAD), a leading advocacy group, attempted to get the FDA to withdraw its approval of cochlear implants for children, based on the premise that deafness isn't a disability that needs to be fixed. In 1998, the NAD withdrew its opposition, and in 2000, the organization adopted a formal position statement on cochlear implants that while far from endorsing cochlear implants, takes a more measured approach.

The statement reads in part:

The NAD recognizes all technological advancements with the potential to foster, enhance, and improve the quality of life of all deaf and hard of hearing persons . . . The role of the cochlear implant in this regard is evolving and will certainly change in the future. Cochlear implants are not appropriate for all deaf and hard of hearing children and adults. Cochlear implantation is a technology that represents a tool to be used in some forms of communication, and not a cure for deafness. Cochlear implants provide sensitive hearing, but do not, by themselves, impart the ability to understand spoken language through listening alone. In addition, they do not guarantee the development of cognition or reduce the benefit of emphasis on parallel visual language and literacy development.

Another section of the statement reads:

The focus of the 2000 NAD position statement on cochlear implants is on preserving and promoting the psychosocial integrity of deaf and hard of hearing children and adults. The adverse effects of inflammatory statements about the deaf population of this country must be addressed. Many within the medical profession continue to view deafness essentially as a disability and an abnormality and believe that deaf and hard of hearing individuals need to be "fixed" by cochlear implants. This pathological view must be challenged and corrected by greater exposure to and interaction with well-adjusted and successful deaf and hard of hearing individuals.

While objections among the deaf community to cochlear implants were stronger during the earlier days of cochlear implantation and seem to have diminished as the numbers of successfully implanted individuals has increased, there is still a sometimes subtle bias against cochlear implants and implantees among some individuals who are deaf.

"The deaf community feels if you choose to be in the hearing world, you are leaving their culture behind," said Ryan McLeod, who received a cochlear implant in 2001, when he was 20 years old, and has realized a significant improvement in his ability to hear as a result. Since being implanted, he has been shunned by some people who are deaf on several occasions. He tells of an encounter with a deaf childhood friend several months after he was implanted. The friend asked McLeod about the implant and told him he would never get one. "I just looked at him, and I was like, 'Why? Ninety percent of the world does hear and 10 percent doesn't. You can be in both worlds. You want to sign, fine, no problem,'" said McLeod. He recalled his friend replying, "I don't need to hear. They have to adjust to me. I don't have to adjust to them."

"I could understand that if 90 percent of the world had a hearing loss and 10 percent didn't, but when the overwhelming percentage of the world hears, I think there has to be some adjustment on your part," said McLeod.

In another instance, McLeod was at a party attended mostly by people who were deaf. Whenever he tried to have a conversation, he had to write down his thoughts. He attempted to speak with others at the party, "but for the most part, they refused to read my lips because they didn't know how, or they knew how and were just like, 'I'm in the deaf culture. You're on my turf, you have to adapt to me.'"

"They feel they have a culture that's unique. Most people don't look at it as, 'I have a disability, I can't hear.' They look at it as, 'Look at me, I'm deaf.

Isn't that cool?' It's a source of pride to them," said McLeod. "Whereas me, I just feel that they let their deafness define who they are. I feel that's a part of me. I have many other interests. And there are many other parts to me other than my hearing loss. For the most part, I feel the deaf community lets their lives revolve around the fact that that they have a hearing loss instead of letting it revolve around the fact that, 'OK, I'm deaf and I have these other things too.' I just look at it as kind of weird."

ONCE THE HURDLES OF RESEARCH, HUMAN TESTING, and regulatory approval are surmounted and neural prostheses become a commonplace treatment for disabilities, as they likely will, the question arises as to whether they can and should be used to enhance the abilities of able-bodied individuals as well. And if brain enhancement implants become available, who should receive them? Who among us would not want the pilot of the airplane you are flying on to have enhanced reflexes or the surgeon performing a complex procedure on you to have a keener sense of sight and more intense powers of concentration than one who was not enhanced?

Many recoil at the thought of someone having a foreign object implanted in his or her perfectly healthy brain, but such procedures might well become culturally acceptable. Consider, for example, that during the seventeenth century, the idea of surgically violating the body was abhorrent. Now, people have elective surgery to improve their looks and no one blinks an eye. And some argue that we already manipulate our brains with alcohol and mood-altering drugs, such as Prozac, Paxil, and Zoloft, not to mention illegal recreational drugs.

Most bioethicists who contemplate the question of enhancement focus on pharmaceuticals, but they do not consider the moral questions implicit in brain implantation to be much different. And surprisingly, some leading bioethicists do not have a problem with the idea of enhancement of the able-bodied.

"I'm not sure whether it will be drugs or devices, but things are coming in both areas," said Arthur Caplan. "What works for people with illnesses will work for people who are normal, most likely to add ability. I don't have a moral problem with that per se. Some argue it's just unnatural to give people capacities more than what they were born with. I think that's exactly what we do in education and training, and libraries, and computers, so I'm not moved by that argument."

But Caplan cautioned that enhancement of the able-bodied must be done on a voluntary basis, without coercion. If a child is implanted with a chip in the brain that will make him or her a good violinist because that is what the child's parents want, that's bad. On the other hand, if one's memory is boosted, or an individual is given enhanced reflexes or an extended sense of smell by means of implantation and can choose whether to use the enhancements, that's fine with Caplan. "To me the right stance is enhancement is acceptable if you can improve and enhance people's abilities, but they still retain the right to reject them." This, he pointed out, is in keeping with what goes on in current culture, in which "many people today don't choose to use the brains they have."

What does concern Caplan about enhancement of the able-bodied is that it has the potential to broaden the existing gap between the haves and have-nots of the world. "It's not the technology; it's the inequity that's the problem. It's the money and access issue," he said.

As for the potential risks involved in implantation, Caplan thinks people should be able to decide for themselves if the risk is worth the reward. But as a practical matter, he feels the medical community will limit the risk: "Strange as it may seem, the moral boundary will probably be set by medicine saying anything more than minimally risky odds can't be taken just for enhancement purposes. I think most responsible doctors won't do it, but there will be some who will. And we will be debating them as renegades and morally questionable people in thirty or forty years."

Thomas H. Murray, president of the Hastings Center, who has written extensively about the ethics of performance-enhancing drugs in sport, has built a complex foundation on which to judge the value of enhancement of healthy people. To his thinking, the notion that risk automatically makes enhancement unacceptable is an argument that doesn't hold up. This, he explained, is because many of those who would benefit from such technologies are already at risk, such as a football player who charges into somebody weighing 350 pounds or a ski racer who hurls downhill at 80 miles per hour. He also rejects the argument that implanting for therapeutic purposes is morally acceptable while implanting for enhancement is not. "If you are imaginative, you can regard almost all therapeutic interventions as enhancements of a sort," said Murray. Antibiotics to fight infections and vaccines to prevent disease, for example, can be regarded as enhancements of one's immune system.

Murray also doesn't go along with the argument that what is natural is good and what is unnatural is bad. This argument fails, he said, because "there are many things that are natural that are not good. Disease is natural, and human cruelty is a manifestation of human nature that is tragically common in this world. If you're going to make natural your standard, you are stuck with a mix of the beautiful and the ugly."

Instead, Murray believes that the way to determine whether biomedical interventions are acceptable is to measure them against the human meaning of a society's practices. Using sport, the area he has given the most thought to, as an example, he pointed out that people generally hold an athlete's talent, hard work, and training in high regard, and not merely his or her level of performance. As a result, "most people seem to think the use of a drug that enhances performance is a kind of undermining or perversion of what we care about in the sport . . . what makes it admirable, beautiful, something we like to watch," said Murray. The same would hold true for a neural prosthetic performance enhancer.

On the other hand, where competition is not an issue, as in the case of an implant that would make a surgeon's hands more steady, an enhancement may well be acceptable, providing that the surgeon is not put at risk by having the implant and is not coerced into having the implant to remain competitive. This is because "the point of the practice of surgery is not to show off the intrinsic physical skill of the surgeon. The point of the practice of surgery is to heal the patient," explained Murray.

In the final analysis, said Murray, "you don't come out with an answer that says, 'Enhancement bad, therapy good.' You come out with an answer that says, 'Therapy good; enhancement, sometimes bad, sometimes good.'"
In the meantime, Murray is still building an analytical framework for making

these decisions based on the human meaning society invests in a practice. Murray concluded by saying, "Till I die, I'll probably still be trying to work out the implications."

He might well have spoken for a lot of the neural prosthetic community.

—◦◦◦◦◦◦—

THE WORLD OF NEURAL PROSTHESES is a relatively small one in which everyone seems to know everyone else, which lends itself in some cases to incestuous research, and not a little bit of competitiveness—not necessarily a bad thing. Despite the fact that these are virtually all brilliant people, most with M.D.'s, Ph.D.'s, or both, they are people, with the same emotions, foibles, and strengths as the rest of us. Though most will not openly deride another's work, they will frequently present reasons why theirs will succeed and the others' won't. Which, of course, is also natural. If they feel their way of doing things will not work, they have no business working on it. That said, as a group, these are brilliant people who for the most part could make a lot more money doing something else but have chosen to focus on neural prosthetic technology not only because it is interesting science but because it has the potential to make lives better for a lot of people.

This motivation is exemplified by John Donoghue, who said, "We've taken something that is rather abstract, understanding computations that form in the brain, in a bunch of cells, which is something that most people at a cocktail party would not quite understand or even be all that interested in, and we can take that information and help people who are in terrible shape with paralysis live better lives. I can't think of anything more fulfilling than that. I'm fortunate in being able to pursue my own interest in understanding computations, but to have it spin off in doing something with this potential benefit is very exciting."

Similarly, Dustin J. Tyler, a biomedical researcher with the Cleveland Veterans Administration who is also on the faculty of the biomedical engineering department at Case Western Reserve University and is affiliated with the Cleveland FES Center, said, "The reason I'm in this field is to help people. I'm not building a widget, I'm building a widget that helps somebody recover function and restore some normalcy in their life." Tyler spoke of the fulfillment he receives from seeing someone with some level of self-sufficiency restored—as when people previously unable to hold eating utensils can feed themselves, or an individual with paralysis has bowel or bladder function restored by an implanted device, or someone who is unable to eat regains the ability to swallow.

Most spinal cord injuries happen to young people in their twenties or thirties. Because medical science has enabled them to live relatively normal lifespans, they face the challenges of paralysis for a long time. Restoring some function for these individuals is, said Tyler, "gratifying. It makes the late hours worthwhile." As for his potential to make more money in the world of commerce, he said, "People are more important than money."

This same sense of fulfillment was voiced by Graham Creasey, a physician also affiliated with the Cleveland FES Center, who specializes in spinal cord injuries: "The whole group of us here is very focused on the patient and whenever we are feeling a little discouraged or tired, hearing a patient say what a difference this makes is something that lifts our spirits and keeps us going."

"It's quite striking even for people who have been in the field many years," Creasey said. "It still can be very moving to hear the stories the patients tell about the difference it makes in their lives. I think what is unique about this is it is possible for the first time in history to restore function that would have been lost for life. Someone who was paralyzed in the past by spinal injury would have been paralyzed for life. It's the first time we've ever been able to move paralyzed limbs and muscles in a useful way. It's very dramatic hearing the effect that has on somebody's life. Not just seeing the limb move but the effect it has on their lifestyle and quality of life. So we are moving much more towards looking at quality of life or those more intangible things as the eventual outcome."

17

Biomimetic and
Superhuman

M edical advancements by their very nature usually move at a glacial pace. Eureka moments are few and far between in the laboratory, and even when they occur, it takes years to move them out into the world of medical practice. Much more frequently, progress in medical technology involves years of research and development and one small step forward at a time, with each step built incrementally on a foundation painstakingly set in place by others over a period of many years. Such is the case in the field of neural prostheses. Yet the pace at which the neural prosthetic glacier is making its way across the landscape of creation is increasing, its inertia building as technological advancement generally escalates and as greater numbers of talented and committed people are attracted to the field. Witness the fact that while physicians, engineers, and scientists have worked on the various elements of neural prosthetic implants for many years, humans have just recently received permanent retinal implants, auditory brainstem implants, motor cortex implants, and functional electrical stimulation implants, albeit on experimental bases. Glaciers move slowly, but the blocks of ice that fall from their leading edges are impressive.

The relevant questions are: What direction will the glacier take in the future, and will its pace continue to increase? The answer, according to Terry Hambrecht, depends in large measure on continuing improvements in implantable electrodes. Specifically, electrodes have to become more selective, and to do that they must be made smaller. Existing electrodes interact with pools of neurons. The theoretical ideal, said Hambrecht, is to "get down to one electrode per neuron. That's not going to happen in our lifetime, but that's the goal, to keep working toward that kind of resolution. Then you'd essentially be able to control the entire nervous system exactly the way you want to control it."

Hambrecht even talked of having multiple electrodes per neuron, giving the ability to separately control the axon, dendrite, and body of a single cell. This could, for example, allow for the detection of a Parkinson's tremor coming down a dendrite and then stopping it before the axon has a chance to distribute it to other neurons. Currently, "we can stimulate neurons that have inhibitory inputs on other neurons," said Hambrecht. "That's the only way we have right now of utilizing neural prostheses to inhibit parts of the nervous system, but that's just in its infancy." Eventually, being able to tune inhibitory stimulation more finely would be advantageous for controlling Parkinson's and other motor diseases, and it would give finer control for other neural prosthetic devices.

If such breakthroughs in electrode development do not occur, then "the future of neural prostheses certainly isn't as great as I can picture it in theory," said Hambrecht. But based on current progress, Hambrecht takes an optimistic view; he feels the electrode barriers will crumble. And he is encouraged by the improvements he sees in the benefits to patients brought about by cochlear implants, the first commercial neural prostheses.

Hambrecht is especially encouraged by the benefits he has seen children derive from these implants. "Kids are amazing with cochlear implants," he said. "They are using the intact portion of the nervous system to learn how to use these devices way better than adults use the intact parts of their nervous systems to learn how to use neural prostheses. A 2-year-old learns language very quickly with a cochlear implant. A deaf child doesn't learn normal language without a cochlear implant. You give adults a cochlear implant, and it's much more of a struggle for them to just replace what they've lost."

Hambrecht is also excited about the potential for learning more about the adaptability, or in medical parlance, the plasticity, of the brain that enables it to adjust itself to work with the input it receives from neural prosthetic devices, however little that may be.

One of his dreams, when working on the development of a visual implant, was to attach an image-processing system driven by a television camera to microelectrodes implanted in the visual cortices of people who are blind. "Not with the promise that they would have normal vision, but to find out how effectively they could use the new input. I wonder how much more information they could get if they were asked to utilize it for mobility rather than just report on what they saw. If you are forced to utilize visual input that comes into the nervous system, you can do a lot more with it," explained Hambrecht. He cited experiments showing that if sighted people look through prisms that show the world upside down, what they see is an inverted world. But if they are forced to navigate while looking through the prisms, their brains adapt so that the world eventually appears right side up.

The goal of the television experiment Hambrecht had in mind was, in his words, to "find out whether or not the rest of the nervous system could help optimize the limited information one will get from early-model visual prostheses." Based on experience with cochlear implants, it may well be able to do so. To find out for sure remains an experiment for the future.

Mark Humayun, an ophthalmologist and retinal implant researcher, pointed out that as the size of neural prosthetic elements, such as electrodes, decreases to the cellular level, the need to make them biocompatible increases. Doing so is the goal of the Center for Biomimetic Microelectronic Systems, which he founded in 2003. Biomimetics refers to implantable devices designed to mimic the human body. The more biomimetic an implant is, the less chance there is for rejection and scarring that can reduce the effectiveness of electrodes.

One way to fool the immune system into thinking an implanted neural prosthesis is a natural part of the body so it does not attack the invader is to place an organic coating over the implanted device. Such coatings may even have the benefit of attracting neurons to electrodes rather than pushing them away. For this to occur, biologists must create coatings that attract glial cells, which constitute the support network of the nervous system. Neurons like to be where glial cells are, so if glial cells are attracted to an implant, the neurons will follow. The benefit of close contact is made clear when one considers that if the electrodes of a retinal implant are as little as 50 microns— the width of two human hairs—away from the retina, the current required to inject a signal doubles from what it would be through direct contact. And the more current required, the greater the chance of tissue damage.

Researchers affiliated with Humayun's Biomimetic Center are working to consolidate neural prosthetic systems, such as the retinal implant, into a single piece of equipment, so that instead of having an array of electrodes attached by wires to a processing unit implanted elsewhere in the body, everything is built into one tiny device. Engineers also hope to do away with the need for external power, which now must be provided to neural implants either by wires running through the skin, radio wave telemetry, or implanted batteries. Batteries eventually need replacement, and as neural prostheses improve and can handle more data, they require more power, making heat dissipation an issue. One answer to the power problem, as Humayun envisions it, is to harness the body's own motion, much as some watches use kinetic energy created by the movement of the wearer's wrist to generate power. For retinal implants, the continuous, rapid (saccadic) movements of the eye—which occur constantly though we are not aware of them—and the rapid eye movements, or REM—which take place while dreaming—are energy sources waiting to be harnessed.

Looking even further into the future, investigators may find that the ultimate biocompatible implant will provide yet further proof your mother was correct when she said, "Eat your spinach, it's good for you." "We've isolated a small molecule that we've gotten from a spinach plant that captures light and releases electrical energy," said Humayun. This has led retinal specialists to wonder whether the molecule could be integrated with cells in the retina to replace damaged rods and cones. If the concept can be made to work, "you would make cells in the eye that were otherwise not light sensitive now light sensitive," said Humayun.

To test this idea, scientists have coated the spinach-derived molecule with liposomes (artificial membranes that resemble cell membranes) and integrated them with ganglion cells in an isolated retina. When light was shone on the molecules, just like Popeye gaining super strength from a can of spinach, the ganglion cells leaped to life. Yet it is a long way from such a laboratory experiment to utilizing the idea in the eyes of people who are blind. One problem is that after a few months the spinach molecules degrade. Humayun's answer to that would be to simply inject a fresh slurry of spinach into the eye. He admitted that the idea "is really far out," though not impossible. Should the concept succeed, the Popeye electrode may take a proud place in the pantheon of stimulating devices.

Another idea that is really far out but not beyond the realm of possibility—should neural prosthetic devices become truly biomimetic and therefore safe—could be dubbed the superhuman concept. "I've never had any qualms about saying that we might make people superhuman someday," said Hambrecht. "We haven't done it yet, but there is no reason, theoretically why we couldn't." By this, he meant it is well within the realm of possibility to enhance any of the human senses. "You could go through every sense and come up with a theoretical way you could extend that sense," he said.

Ironically, those who are outfitted with electrodes in their brains to overcome sensory disabilities, such as blindness or deafness, may be the first superhumans. If electrodes are already in place in their eyes, brains, or ears, Hambrecht sees no reason why sensors capable of capturing signals from colors and sounds beyond the range of normal human perception couldn't be connected to those electrodes. Thus, the formerly blind could see into the ultraviolet and infrared fields and other parts of the color spectrum, and the heretofore deaf could hear in the range above 20,000 hertz, like many animals.

Similarly, researchers have been able to evoke vivid memories that were otherwise seemingly irretrievable in patients by experimentally stimulating certain parts of the brain. If this process becomes better understood and harnessed, it could have potential for Alzheimer's patients as well as for increasing the memory capacity of individuals who are not impaired.

Admittedly, such innovations will not take place for many, many years to come, if ever, but today's neural prosthetic science brings them into the domain of the possible.

As for the hopes of those who have already offered their bodies up for the testing of neural prosthetic devices, they are considerably more modest.

As Jennifer French, who has received a standing system (see chapter 7), so eloquently put it, "My outlook in terms of what is going to happen with spinal cord injuries is, we are not going to have a magic bullet I'm going to be able to take and walk out of my wheelchair, and there is not going to be some magic surgery after which I'll suddenly walk out of the hospital bed. It's going to be a merger between a pharmacological or surgical solution and a therapeutic solution, and that therapeutic solution, I believe, is going to be somewhere in neural prostheses. They have to merge those solutions together in order to come up with a real solution for spinal cord injury . . . A lot of the scientific community is moving toward that view, which is fantastic."

Selected Bibliography

Agnew, William F., and Douglas B. McCreery, eds. *Neural Prostheses: Fundamental Studies*. Englewood Cliffs, N.J.: Prentice Hall, 1990.

Bilger, R. C. "Evaluation of Subjects Fitted with Implanted Auditory Prostheses." *Annals of Otology, Rhinology, and Laryngology* 86, suppl. 38, no. 3, pt. 2 (1977).

Brazier, Mary Agnes Burniston. *A History of Neurophysiology in the 17th and 18th Centuries: From Concept to Experiment*. New York, N.Y.: Raven Press, 1984.

Delgado, José M. *Physical Control of the Mind: Toward a Psychocivilized Society*. New York, N.Y.: Harper & Row, 1969.

Dupont Salter, Anne-Caroline, Stephen D. Bagg, Janet L. Creasy, Carlo Romano, Delia Romano, Frances J. R. Richmond, and Gerald E. Loeb. "First Clinical Experience with BION Implants for Therapeutic Electrical Stimulation." *Neuromodulation* 7 (January 2004): 38.

Hambrecht, F. Terry, and James B. Reswick. *Functional Electrical Stimulation: Applications in Neural Prostheses*. New York, N.Y.: Marcel Dekker, 1977.

Himrich, Brenda L., and Stew Thornley. *Electrifying Medicine: How Electricity Sparked a Medical Revolution*. Minneapolis, Minn.: Lerner Publications, 1995.

Martin, Richard. "Mind Control." *Wired*, March 2005, 114–19.

Pera, Marcello. *The Ambiguous Frog: The Galvani-Volta Controversy on Animal Electricity*. Translated by Jonathan Mandelbaum. Princeton, N.J.: Princeton University Press, 1992.

Purves, William K., David Sadava, Gordon H. Orians, and H. Craig Heller. *Life: The Science of Biology*. 6th ed. Gordonsville, Va.: W. H. Freeman and Company; Sunderland, Mass.: Sinauer Associates, Inc., 2001.

Rowbottom, Margaret, and Charles Susskind. *Electricity and Medicine: History of Their Interaction*. San Francisco, Calif.: San Francisco Press, 1984.

Schmidt, E. M., M. J. Bak, F. T. Hambrecht, C. V. Kufta, D. K. O'Rourke, and P. Vallabhanath. "Feasibility of a Visual Prosthesis for the Blind Based on Intracortical Microstimulation of the Visual Cortex." *Brain* 119 (April 1996): 507–22.

Thompson, Richard F. *James Olds, 1922–1976: A Biographical Memoir*. Washington, D.C.: National Academy Press, 1999.

Yeomans, John Stanton. *Principles of Brain Stimulation*. New York, N.Y.: Oxford University Press, 1990.

Index

action potentials, 5, 29, 65, 154, 206, 245, 247
Advanced Bionics, 170–171, 204
amyotrophic lateral sclerosis (ALS), 5, 216, 221
Andersen, Richard A., 224–225
audiologist, 14, 129, 144–146
auditory brainstem, 12, 149, 188, 256
auditory brainstem implant (ABI), 132–141, 144–146, 148, 150, 185, 187, 190, 273. See also Davidson, Marilyn; penetrating auditory brainstem implant

Baily, Mary Ann, 260
balance, the sense of, 31, 96, 106, 111, 113, 129, 162, 187, 236–237, 263. See also vestibular system
Bardeen, John, 51
Bartholow, Roberts, 42–43
Basmajian, John, 77
battery, 10, 33, 39, 95, 110, 129, 181
battery-powered systems, 204, 205
Bell Laboratories, 51–52, 191–192
Belmont Report, 257–258
Benabid, Alim-Louis, 237
Berger, Hans, 227
Berger, Theodore W., 244–250
Bilger, Robert, 17
Bilger Report, 17–18
biocompatible, 33, 48, 54, 171, 275–276
bioengineering, 179, 184, 195, 197, 253

biomedical engineering, 75–76, 79, 82, 104–105, 160, 191, 199, 244, 271
BION, 198–199, 200, 202–209, 279
Bionic Technologies, 213, 261
bladder, 55, 66, 95–96, 118–120, 122–123, 209, 272
bladder and bowel implant. See Vocare
Brackmann, Derald, 18, 144–145
Braille Institute of America, 157
brain: beta and mu rhythms, 228–229; brain tumor, 129; brain waves, 1, 139, 222–223, 225, 227–9, 232; cerebellum, 46; cerebral cortex, 30, 42–43, 65, 218–219; cochlear nucleus, 133–134, 139–140, 148, 150, 187, 190; enhancement of, 268–270; globus pallidus, 237; hippocampus, 244–250; imaging of, 242; inferior cerebellar peduncle, 135–136; inferior colliculus, 150; language of the, 238; lateral geniculate nucleus, 154; occipetal lobe, 45; posterior parietal cortex, 224–225; subthalamic nucleus, 237; thalamus, 43, 185, 200, 237, 239; visual cortex, 29, 203, 224. See also auditory brainstem; motor cortex
Brain Communicator, 222–223
brain-computer interface, 197, 201, 211, 213, 222–223, 227–229, 231–235
BrainGate, 197, 213–217, 219
Brattain, Walter, 51–52

Brindley, Giles, 49, 55–69, 72, 118–125, 196. *See also* motor prostheses; visual cortex implants; Vocare
Brown, Molly, 141–148, 156, 211, 214, 218–220
Brown University, 156, 211, 214, 218
Brummer, Barry, 186, 189
Button, J., 48

California Institute of Technology, 224
Caplan, Arthur, 261–262, 268–269
Case Western Reserve University, 64, 74–77, 81–84, 104–105, 122, 228, 263, 271. *See also* Cleveland FES Center
Caviness, Verne, 217, 218
Center for Bioethics, 252
Center for Biomimetic Microelectronic Systems, 199, 248, 275
Center for Neural Communication Technology, 193
Center for Neural Engineering, 244, 248
Center for Neural Interfaces, 195, 213
Chow, Alan, 172–176, 179–181
Chow, Vincent, 172–173
Churchey, Harold, 161–168, 171–172
Cleveland Functional Electrical Stimulation (FES) Center, 64–65, 82, 84–85, 91, 103–105, 111, 114–115, 118, 121–122, 263, 271–272
cochlea, 2, 12–13, 17–19, 21–24, 32, 47–48, 132–133, 139, 183, 190. *See also* ear: inner
cochlear implants, 3, 143–144, 146, 148, 151, 155, 164–165, 169, 254; in children, 274–275; development of, 16–24, 47–48, 131–132; electrodes for, 17–25; ethical controversy, 265–268. *See also* Eddington, Donald K.; McLeod, Ryan; National Association for the Deaf; Pierschalla, Michael
cochlear implants, manufacturers of: Clarion, 170–171, 202, 204; Cochlear

Corporation, 134, 139; Ineraid, 19–20; Med-El, 26
cognitive implant, 244–251
Colletti, Vittorio, 148–149
Columbia University, 264
computer: controlled, 68, 77; cursor, 216–217, 221–222, 224–225, 226–231, 234; generated, 21, 68, 248–249; modeling, 85, 247, 250; programming, 22, 62, 80, 222, 238. *See also* brain-computer interface
control systems, 4, 71, 91, 203; implantable joint angle transducer, 92–93; implantable stimulator telemeter, 94–95; joystick, 5, 89, 91, 93, 116; myoelectric, 93, 95, 105
Creasey, Graham, 118–124, 126, 272
Cyberkinetics Neurotechnology Systems, 213, 215–216, 221, 261
Cyberonics, 240, 265

Davidson, Marilyn, 128–138
Davis, Michael, 259, 262
deep brain stimulation, 236–243
Defense Advanced Research Projects Administration, 260
de Forest, Lee, 50
de Juan, Eugene, 159–175
Delgado, José, 46–47, 279
depression, 10, 14–15, 21, 136, 239, 241–242
Djourno, Andre, 47
Dobelle, William, 200–201
Doheny Eye Institute, 152, 156–157, 164
Donoghue, John P., 197, 211–214, 217–221, 223–224, 261, 271
Duchenne, Guillaume Benjamin Amand, 41–42
Duke University, 71, 158, 167

ear: inner, 2–3, 12, 115, 128, 138, 183; middle, 12; outer, 12

Eddington, Donald K., 17–21
Edison, Thomas, 49
electrical signals, 1–2, 12, 25, 30, 32, 40,
   65, 70, 80, 154, 191, 222, 228. *See also*
   action potentials
electrical stimulation, 48, 107, 159, 163,
   169, 173, 194, 202, 208, 237, 239–240,
   243; by-products of, 186–187; controlling behavior by, 45–46; control
   motion by, 64, 74, 76–77, 81–83,
   86; safe levels of, 59, 135–136, 161–
   162
electricity, 4, 70–71, 77, 87, 92, 103, 109,
   116, 121, 139, 162, 174, 202, amplitude,
   186–187; charge, 28, 31, 41, 49–50,
   36–37, 111, 114, 163, 187–189, 233,
   236: current, 18, 20
electricity: in animals, 44, 46; in the
   brain, 196, 201, 239, 253; flow of, 114;
   in muscles, 39, 59–60, 71, 90, 94, 198,
   205, 208; in nerves, 31–32; in the retina, 169, 174; in wire, 41–42
electrocorticography, 233–234
electrodes, 32, 41–49, 51, 196–198, 200–
   205, 252, 256, 261, 265, 274–277; in
   the brain, 211–214, 216, 218–223; corrosion of, 32, 189, 190; drifting of,
   140; ECOG, 233–234; epimysial, 108;
   implantable, 83, 89, 228, 230, 232,
   274; materials, 59, 87, 89, 108, 135,
   152, 162, 186–187, 197; microelectrodes, 5, 53, 62, 73, 140, 185,
   188–191, 194, 220, 253, 275; nerve
   cuff, 114–115; neurotropic, 221, 223;
   noninvasive, 222, 226–229, 231–235;
   penetrating, 141, 145, 148–150, 196;
   platinum, 87, 89, 152; recording, 3,
   65, 91, 94, 96, 191–193, 201–202,
   220–226, 231–234, 247–249; surface,
   71, 140–141, 148–149, 183, 188, 196,
   253
electroencephalogram, 222

electroencephalography, 222, 227–228,
   230–234
electromyographic, 96, 223
electronics, 1, 14, 25, 36, 54, 160, 167–
   168, 171, 192, 202–206
Engineering Research Center for Wireless Integrated Microsystems, 191
England, 50, 55, 62, 98–99, 121–123,
   216. *See also* Brindley, Giles; Creasey,
   Graham
Environmental Impact Center, 186, 189
epilepsy, 46, 194, 233, 239, 240, 265. *See
   also* deep brain stimulation
erectile dysfunction, 55, 67, 121
ethics, 252, 256–257, 259–261, 263–
   264, 269
exercise, 82–83, 103–104, 110–112, 147,
   153, 160, 207, 230, 263
Eyries, Charles, 47

Faraday, Michael, 41, 47
FineTech Medical Ltd., 66, 118, 120. *See
   also* Vocare
Fleming, John Ambrose, 50
Foerster, Otto, 44–46, 58–60
Food and Drug Administration (FDA),
   53, 62, 77, 158, 169, 201, 257–258,
   262, 264–266; device approval, 4, 26,
   66, 90–91, 94, 112, 121, 118, 139, 150,
   180, 213, 216, 238, 240–241
Frank, Karl, 61, 72, 83, 185
Franklin, Benjamin, 28, 38, 70
Freehand, 4, 5, 85, 89–93, 97, 103–105.
   See also Cleveland FES Center; func-
   tional electrical stimulation; Jatich,
   Jim; Peckham, P. Hunter
French, Jennifer S., 37, 41, 47–48, 68,
   98–104, 106–114, 116, 237, 278
Friehs, Gerhard, 214
Fritsch, Gustav Theodor, 42
functional electrical stimulation (FES),
   4–5, 74, 198, 202, 223, 231, 273;

functional electrical stimulation (*cont.*)
development of, 81–85; for balance,
115–116; for bladder and bowel control, 117–127; for hand grasp, 86–97;
for standing and walking, 64, 103–
114. *See also* Cleveland FES Center;
French, Jennifer; Hambrecht, F. Terry;
Peckham, P. Hunter; Triolo, Ronald J.

Galvani, Luigi, 38–40
Georgopoulos, Apostolos, 223
Gilbert, William, 35, 75
Glenn, William, 47
Goodman, Robert, 264–265
Greenberg, Robert, 158, 165, 167–171, 255

hair cells: 2, 12–13, 17, 19, 32. *See also*
ear: inner; hearing
Hambrecht, F. Terry, 70–71, 79–80, 83;
on the Bilger Report, 17; on Brindley,
Giles, 62; on electrodes, 185, 187–188,
194, 196, 201–203; on the future, 273–
275, 277, 279; and the NIH Neural
Prosthesis Program, 61, 72–75; on
Michael Pierschalla, 24; on visual
prosthesis, 253–255; *See also* National
Institutes of Health; Neural Prosthesis
Program
Hamel, Scott, 226–230
Harvard College, 218, 244, 246–247;
medical school, 20
Hastings Center, 259–260, 269. *See also*
Baily, Mary Ann; Murray, Thomas H.
hearing, 3–4, 39, 185; mechanical, 137,
146; normal, 22, 137, 266. *See also*
auditory brainstem implant; cochlear
implants; ear
hearing aid, 14–15, 17, 21, 23, 129
Heath, Robert, 45–46
Heetderks, William, 73, 202–203, 215–
216, 231–232, 257
Hertz, Heinrich, 50

Hess, Walter Rudolf, 43
Hippocratic Oath, 256
Hitselberger, William E., 132–133
Hitzig, Julius Eduard, 42
Hoffman, Sharona, 263–264
House, William, 17, 129, 131–134
House Ear Institute, 17–18, 22, 132–133,
139, 140–141, 144, 185, 187. *See also*
House, William; Otologic Medical
Group
Humayun, Mark, 157–165, 167–168,
170–171, 173–175, 199, 248, 275–277.
See also Center for Biomimetic Micro-
electronic Systems; retinal implants
Huntington Medical Research Institutes,
135, 140, 149, 184–188, 190. *See also*
McCreery, Douglas B.; penetrating
auditory brainstem implant

Illinois Institute of Technology, 190, 203,
253, 259, 262
implants. *See* auditory brainstem;
cochlear; cognitive; deep brain stimulation; functional electrical stimulation; motor cortex; penetrating auditory brainstem; retinal; standing;
visual cortex; walking
informed consent, 253, 256–259, 263–
264. *See also* Belmont Report
institutional review boards (IRBs), 257,
262–264
insurance, 66, 102–103, 147, 260
integrated circuit, 49, 53, 192, 205. *See
also* transistor, photolithography

Jatich, Jim, 85–89, 91–97. *See also* Freehand; Cleveland FES Center; Peckham,
P. Hunter
Johns Hopkins University, 159, 167–168,
171, 180, 200; school of medicine, 58,
71. *See also* Wilmer Eye Institute
Johnson, Gloria, 175–177

Kennedy, Philip R., 217, 221–223
Kilby, Jack, 52
Koester, Holly S., 117–126

Lenarz, Minu, 150
Lenarz, Thomas, 150
Lewin, Walpole, 49, 55, 60
Leyden jar, 28–29, 36, 38–39
Loeb, Gerald E., 199–205, 208, 279
Lou Gehrig's disease. See amyotrophic
    lateral sclerosis
lower extremity control unit, 110, 112
Lozano, Andres, 238–239, 241–242

macular degeneration, 4, 158–159
Mann, Alfred E., 169–170; Foundation,
    199, 203, 204; Institute for Biomedical
    Engineering, 199
Marconi, Guglielmo, 50
Marshall, Mary Faith, 252, 261
Marsolais, E. Byron, 105
Massachusetts Eye and Ear Infirmary, 17,
    19–20; Cochlear Implant Research
    Laboratory, 17, 20. See also Eddington,
    Donald K.
Massachusetts Institute of Technology,
    19, 71, 253
Mayberg, Helen, 241
McCreery, Douglas B., 184–190, 279
McFarland, Dennis, 229, 231
McLeod, Ryan, 267–268
Medical Research Council. See United
    Kingdom's Medical Research Council
Medtronic: Activa, 238, 240; Intercept
    Epilepsy Control System, 238, 240
Merton, Patrick, 56
MetroHealth Medical Center, 84, 263
microelectronics, 198–200
Mortimer, J. Thomas, 74–84, 122–123
motor cortex, 2, 29, 42, 59–60, 65
motor cortex implants, 91, 197, 202,
    212–213, 215, 218, 221, 223–224, 228,

232–233, 273. See also BrainGate,
    Donoghue, John; Nagle, Matthew;
    Normann, Richard; Schwartz,
    Andrew
motor prostheses, 63–65, 73, 77–79, 91.
    See also Cleveland FES Center; Free-
    hand; French, Jennifer; functional
    electrical stimulation; Jatich, Jim;
    Peckham, P. Hunter; Triolo, Ronald J.;
    standing implants; walking implants
mu rhythms, 228–229
Murray, Thomas H., 192, 269, 270–271
Muscle Communicator, 222–223
muscles: bladder and sphincter, 122–
    126; electrical stimulation of, 5, 32,
    42, 64, 111, 114–115, 198, 204–208;
    fatigue, 80; implanting, 108–110;
    involuntary twitching of, 134–135;
    paralysis of, 5, 70–71, 76–77, 81–83,
    87, 103, 202; recording from, 78, 94,
    105, 201; types of, 80–81
Musschenbroek, Pieter van, 36
myelin, 130

Nagle, Matthew, 213–216, 218, 221, 224
National Association for the Deaf (NAD),
    266
National Commission for the Protection
    of Human Subjects of Biomedical and
    Behavioral Research, 257
National Institute of Biomedical Imaging
    and Bioengineering, 215, 231
National Institutes of Health (NIH), 17,
    61, 72, 195, 201, 219. See also
    Hambrecht, F. Terry; Heetderks,
    William; National Institute of Bio-
    medical Imaging and Bioengineering;
    Neural Prosthesis Program
National Research Act, 257
National Science Foundation, 197, 260
National Spinal Cord Injury Statistical
    Center, 90

nerves: auditory, 12, 17, 19, 22, 48, 129–
133, 136, 142–145, 148–150; facial,
130, 132–133, 138, 144; nerve fibers,
12, 17, 66, 81, 114–115, 121, 140, 154,
254; optic, 32, 40, 154, 174, 196, 254;
vestibulocochlear, 135. See also neurons
nervous system, 49, 51, 53, 217; makeup
of, 1–3, 7, 30–31, 33, 130–131, 202,
240; implants and the, 41, 45, 72, 95,
103, 185, 191–192, 237, 274–275. See
also brain; nerves; neurons
Neural Prosthesis Program, 185, 202,
215, 231; formation of, 61–62, 72–74;
funding of programs, 24, 83, 170, 188,
192–193; visual prosthesis project,
253–254; workshop, 26. See also
Hambrecht, F. Terry; Heetdeerks,
William; National Institutes of Health;
Neurological Prosthesis Unit of the
United Kingdom's Medical Research
Council
neural prosthetics: funding of, 24, 66,
73–74, 83, 123, 196, 203, 221, 260–
262. See also cochlear implants; cognitive implant; deep brain stimulation;
functional electrical stimulation (FES);
retinal implants; standing implants;
walking implants
neurofibromatosis type 2 (NF2), 130–
131, 139, 143, 148–150
Neurological Prosthesis Unit of the
United Kingdom's Medical Research
Council, 56, 61, 64, 66. See also
Brindley, Giles
neurons, 29, 238, 245, 248–249, 274;
axons, 30; dendrites, 30; firing of, 215,
228; synaptic connections, 30–31, 154,
228, 244. See also nerves
Neuropace: Responsive Neurostimulator,
240
neuroscience, 192, 195, 211, 218, 224,
244, 247

Neurotech Network, 113
neurotransmitters, 12, 29–30, 243
neurotropic, 179, 221, 223
New York Times, 52
Nollet, Jean-Antoine, 37–38
Normann, Richard A., 194–197, 213–214,
220, 261. See also Utah electrode array
Noyce, Robert, 52
Nuremberg Code, 256

Oersted, Hans Christian, 41
Office of Naval Research, 250
Olds, James, 43–45, 279
ophthalmic, 60, 152, 160, 164–165, 168,
173
ophthalmologist, 11, 161, 172, 181, 275
ophthalmology, 158–159, 173, 195
Optobionics, 172, 174, 176
Otologic Medical Group, 17, 129. See also
House Ear Institute
Otto, Steve, 144

paralysis, causes of, 2, 99, 132–133;
organizations for the paralyzed, 90,
123; research into, 28, 38; treatment
of, 4–5, 28–29, 37, 40, 42, 55, 63–66,
74, 72, 197, 213, 215–216, 271. See also
BIONs; Brindley, Giles; Creasey, Graham; Cleveland FES Center; French,
Jennifer; functional electrical stimulation, Hambrecht, F. Terry; Jatich, Jim;
Koester, Holly; Mortimer, J. Thomas;
Peckham, P. Hunter; Triolo, Ronald J.
Pacesetter Systems, 170
Paralyzed Veterans of America, 123
Peckham, P. Hunter, 72, 81–85, 87–92,
96, 105, 122. See also Cleveland FES
Center, Freehand
penetrating auditory brainstem implant
(PABI), 140–141, 144–145, 147, 148–
150, 185, 189–190. See also auditory
brainstem implant; Brown, Molly

Penfield, Wilder, 44–46, 58, 60
percutaneous connector, 47, 105, 134, 138
photolithography, 191, 193. *See also* integrated circuit; transistor
photoreceptors, 2, 32, 56, 154–155, 160–161, 163, 169, 173–173, 179–180, 188, 195, 276
physical therapy, 83, 102, 106–107, 207
Pierschalla, Michael, 8–17, 20–22, 24–27, 41, 259, 262
Pollak, Pierre, 237
processor, 23–26, 105, 109, 114–115, 134–135, 138, 140–141, 145–146; multichannel, 22; portable, 18–19, 21, 166
Pudenz, Robert, 186
Putnam, T., 48

radio, 10, 49–50, 52, 60, 77, 152, 212, 276; radio frequency, 23, 47, 89, 109, 198, 203, 205, 208; transmitter, 77, 152; waves, 47, 50, 90, 138, 204
Rainge, Ronnie, 175–181
Ray, John, 221–222
receiver-stimulator, 194; in auditory brainstem implant, 138, 144, 152; in bladder implant, 120, 125; in cochlear implant, 23; in hand manipulation system, 89, 91–92; in retinal implant, 164–166, 171; in standing system, 107–109, 111
receptors: tactile, 2, 4, 5–6, 32, 103, 224. *See also* photoreceptors
Reswick, Jim, 76–77, 79, 279
retina, 3, 32; functioning of, 56–57, 154–155, 169, 179, 276. *See also* retinal implants
retinal implants, 3–4, 273; development of, 160–166, 171, 173–174, 275–276; functioning of, 151–155, 157–159, 166–167, 175–182, 199. *See also* Chow,

Alan; Churchey, Harold; Humayun, Mark; Greenberg, Robert; Optobionics, Rainge, Ronnie; Schoeman, Connie; Second Sight Medical Products; Zaccaro, Maria
retinitis pigmentosa, 4, 154–158, 161, 173, 175, 179, 181–182
Robblee, Lois, 189
robot, 5–6, 105, 202, 212, 223, 230–232
Rochester Institute of Technology: National Institute for the Deaf, 11, 14; School of American Crafts, 11, 14, 27
Roy, Arup, 152–153
Rushton, David, 64

saccadic eye movements, 171, 196, 276
Schoeman, Connie, 152–158, 167, 171–172
Schulman, Joseph, 170, 199–200, 203–204
Schwartz, Andrew, 7, 223
Second Sight Medical Products, 158, 167, 170–172, 174, 255
sexual stimulation, 43, 66–67, 119, 121
Shannon, Robert V., 22–23, 132–133, 138–142, 148–150
Shealy, Norman, 79
Shockley, William, 51
signal-processing, 24, 198, 229; continuous interleaved sampling (CIS), 25; spectral peak signal (SPEAK), 25
Simmons, Blair, 48
somatosensory, 2, 5, 201–202
speech-recognition, 141, 250
spinal cord, 30, 108, 112, 119, 217; functioning of, 78–79, 98–99, 201–202, 233; injuries to, 5, 66, 90, 101–102, 105–107, 117, 125, 206, 213, 215, 231, 272, 278; research involving, 39, 72; sacral section, 120; spinal root, 79, 120; tumors on, 130. *See also* paralysis

standing implants, 63, 96, 104, 106, 108, 110–111, 115–116, 278. See also French, Jennifer; Triolo, Ronald J.
Stanford University, 48, 191, 192
stimulation. See Cleveland FES Center; deep brain; electrical; functional electrical; muscles
Stritch School of Medicine, 172
superhuman, 101, 277
Surgenor, Timothy R., 216
surgery, 118, 149, 163, 175, 208, 230, 233, 256, 264, 270, 278; auditory brainstem implant, 132, 139–140, 144, 149; awake during, 45–46, 159, 233, 242; brain implant, 214; corrective, 118; deep brain implant, 237–239; following, 20, 94, 106, 108–109, 115, 129, 134, 141, 145, 157, 177–178, 221–222; Freehand implant, 89; retinal implant, 157, 164–165, 179, 181; standing implant, 96, 108–109, 111–112; translabyrinthine procedure, 131, 133; visual cortex implant, 60
Sylmar, California, 158, 167, 169–170

tactile, 2, 4–6, 32, 42, 103, 201, 206, 224
temporal bone, 133
therapeutic effect, 175, 179
Thomson, Joseph John, 49–50
transcutaneous, 19, 138
transistor, 34, 49, 51–53, 79, 191
transmitting coil, 109–10, 120, 198, 205
Triolo, Ronald J., 104–107, 112–115
Troyk, Philip R., 203–204, 253–256, 259
Tuskegee Syphilis Study, 251, 257
Tyler, Dustin J., 271–272

United Kingdom's Medical Research Council, 61, 64; Neurological Prosthesis Unit, 56, 65. See also Brindley, Giles

University of California at San Francisco, 170, 202
University of Michigan, 150, 191–192. See also Center for Neural Communication Technology
University of Minnesota, 185, 223. See also Center for Bioethics
University of Pittsburgh School of Medicine, 223
University of Southern California, 152, 156, 164, 199. See also Center for Biomimetic Microelectronic Systems
University of Tübingen, Germany, 235
University of Utah, 17–19, 201. See also Center for Neural Interfaces
urinary tract infection (UTI), 66, 112, 123–124
U.S. Public Health Service, 72, 251
Utah electrode array, 197, 213

Valencia, California, 169, 199
Verona, Italy, 148
vestibular system, 13, 115–116, 128, 130–131. See also balance, the sense of
Veterans Administration, 118, 180, 260, 271
vision, 45, 145, 153–157, 161, 163, 170, 172–175, 177–182, 195–196, 254, 275; color, 157. See also brain: occipital lobe, visual cortex; macular degeneration; photoreceptors; retina; retinal implants; retinitis pigmentosa
visual cortex implants, 48–49, 55, 119, 190–190, 195–196, 200–201, 203, 253–255; electrodes for, 59–63, 72. See also Brindley, Giles; Dobelle, William; Hambrecht, F. Terry; Troyk, Philip
Vocare bladder and bowel control system, 4, 66, 117–121, 124–126. See also

Brindley, Giles; Creasey, Graham;
   Koester, Holly
Vodovnik, Lojze, 76–78
Volta, Alessandro, 39–40
von Guericke, Otto, 35–36

Wadsworth Center, New York State
   Department of Health's, 228
walking implants, 96, 105
Walter, W. Grey, 43, 46, 51
Williams, Sam, 170

Willowbrook State School, 251; hepatitis
   study, 257
Wilmer Eye Institute, 161, 163, 173, 175.
   See also Chow, Alan; de Juan, Eugene;
   Humayun, Mark
Wired magazine, 214, 216, 279
Wise, Kensall D. 191–194
Wolpaw, Jonathan R., 228–235

Zaccaro, Maria, 181–182